THOUGHTSCAPE
WORD
DROPS

BETSYBOO

Ordering Information:

Prime Seven Media
518 Landmann St.
Tomah City, WI 54660

Printed in the United States of America

Table of Contents

THOUGHTSCAPE WORD DROPS: WONDER OF OUR WORDS

Let us put our war of words to good use: *LIFE is for LIVING, GROW in MIND, BODY, and SOUL* **The Beginning, The End, The First, The Last.** But in the future—or in the death of our words—those words will carry weight for years to come.

The Shadows in Life Keep Following One

OUR WORDS, ACTION, WE CAN THINK ABOUT ANYTHING WE WANT BUT OUR THINKING CAN always be changed for better reasons.

Do not stand behind me; I may not lead Do not stand before me; I may not follow Instead, stand beside me and be my friend Life is for living and growing in mind, body, and soul. Hope—the ability to see a flicker of light amid all this darkness—is a beacon of reassurance.

It has been enlightening to examine words in Bible verses. It's so interesting—it even heals the body. This book outlines key parts from my earlier works. Let us put our war of words to good use, for better or for worse.

Let us focus on the survival of human life and this planet

Preface

DO OUR WORDS AND REASONING cause life or death? How can they be used in fairness to all humanity and cultures?

I have always wondered if there is more to life than what we believe to be "living." Did I know that Bible verses could help us understand the true reason for our existence? Looking at the bigger picture, both ahead and beyond, my path in life eventually turned in a specific direction.

It wasn't planned. As a teenager, like most people my age, I went out and about with friends. One night, my life took a drastic turn—I became a rape victim. This tragedy changed the course of my plans immensely.

I was married at 16 for eight years, and my first child was born in 1962. Three more children followed. In 1970, I lost my late husband to drowning. While raising four children alone, I often wondered why we were created and what our purpose was.

I saw friends and family running around, saying and doing the same things against each other, getting nowhere, and repeating the same mistakes. As my wonder grew, I realized there had to be more to life than what we were experiencing.

I grew up hearing many wise old sayings. What was the reasoning behind them? There just had to be a purpose. Through a lot of trial and error over the years, I began piecing it all together. I realized that the repetition of words and patterns in life would happen over the next generation.

The same mistakes would occur time and again unless corrected. Messages of wisdom and unspoken words came to me. I put the pieces together in my family tree and in my mind. The more I mused, the more I thought there had to be a way.

As I said before, I may repeat sentences in chapters, but that's okay. Each time, I improve or add to the sentence. If you recognize part of this book from your family or relations, it is essential to write down the date, month, day, and hour. Keep a record of your knowledge—it places you on a level with others in your family, even if they never met one another. The wisdom of words still flows through different family trees.

This book shows a way to use words, inspired by Bible verses. It explains why I've written it and why I felt compelled to put pen to paper. I discuss my theories, insights, visions, and wonder since birth. I touch on nursery rhymes, old-time sayings, good and bad words, and their meanings. I show how every word reflects our actions through life and holds significance.

We all inherit parts of Bible verses through our family trees. By putting pen to paper, I became aware of the path we should follow.

Life is about more than growing up, getting married, having children, retiring, and dying.

I have repeated some chapters from my first two books: *Life's Most Complex Problems* (author: Betsy Boo) and *Life's Biggest Jigsaw Puzzle* (author: Elizabeth Campbell Cargill). Each book leads to the next for improvement. This book, *Thoughtscape*, written under the name Betsy Boo, builds on those works.

It has been enlightening to examine words in Bible verses. It's so interesting—it even heals the body. This book outlines key parts from my earlier works. Let us put our war of words to good use, for better or for worse.

Let us focus on the survival of human life and this planet.

The Beginning, The End.

CHAPTER 1

THE ROLLING CLOUDS GATHER NO MOSS

*M*ANY HANDS MAKE LIGHT WORK. as do rolling words. Cleaning our bodies, not our minds, will not solve the world's weather problems. Any book that is written will always attract one attention, like the verses in the Bible. There is always something in what is written in one mind that one would want to know more about. They say the answers are all in the Bible. This is true when we break our words down. Then, one matches them with verses in the bible in numerology and the alphabet. No matter what one reads, there will always be something that enlightens one, whether it is the past or the present. An Asian friend mentioned their country was working on it, it rains at night and fine during the day. My theories, vision, and wonder as a child—insight into what was out there. Also, when one speaks a sentence, it will always stretch further in a long conversation which gathers around our waist. This book explains how we can understand the meaning of our words and how the bible does have all the answers. Especially when it comes to balancing out the weather. Every word in all cultures forms clouds, creating the weather. In other words, love others as I love you. True love comes from the heart, not the mouth. Give without wanting. Give what one needs at any given time. Not what one wants. Our unspoken words, like a rolling stone, gather no moss. It all ends up in cloud formations. The reason behind putting pen to paper in a book form is how we can define our words and understand the verses in the bible they match: alphabet numerology and Technology. It will get us around the world fast with the balance of the weather as we go. as shown in this book of words. This is a theory, insight, wisdom, vision, and wonder.

They talk about the sea levels and the melting of ice, which could flood the earth. But if we have the rotation Balance of the weather around all the planets nearby, Mars, Venus, Mercury, and Earth, With our family trees, words, wisdom, and insight, would it be possible to bring down the tension, anger, and energy force that surround each planet and thus close to a person to unravel down the family trees then around each planet when sea levels rise on earth it spreads around these other planets? Even out of the rotation between each planet with our words, it is possible that we had it all before, and the ice, snow and frozen seas all came down to earth as we spin around the globe the universe. All cultures worldwide work on it. We could prevent the worst from happening. So we don't just flood Earth. It goes around these other planets. Once we work on it, in all cultures, it will work. Please open the door to each earth; let us have something to live for, not to die for. It could work to unravel wise wisdom words between us and these other planets. All cultures on all levels. Men, women, ICE, SEAS, SNOW. == I S, S INNER SUNSHINE

How to live a long life by unravelling words, good or bad, and matching them up with the verse in any bible uses. Start with the dictionary- The dictionary is all about words and meaning. As we all know, it tells us what an individual word means. I asked a school friend what the word repertoire meant. His reply was to look in the dictionary, so I did. My dad had a racehorse called Repertoire. I was amazed to learn it meant repetition. We all get parts of the bible on us over years of readings. By breaking down words that block our view, past and present, we will understand the messages from the almighty. USE OUR WORDS to balance the weather. Take We lived before, made a mistake, and broke the universe apart four times; each time, we had to start from scratch again. Hopefully, people will notice how meaningful our lives are this time and help balance out weather patterns, Mother Nature, and the level in one's family tree first. Then, our words will be used to advance weather patterns.

. AND SALT. T. S= TIME SUN Using a single letter of words represents the weather or wisdom for the next level. Seas =(7) I saw four angels standing at four corners of the earth holding back four winds. And the ocean became smooth as glass. And another angel is. Carrying The great seal of the living god. ICE-17-(8) GOD'S NUMBER. SNOW-71-8 (8) ANOTHER 8 SEAS IS SEVEN,

THE AXLE

axle = 44-(8) When the lamb broke the seventh seal, there was silence throughout heaven, and I saw the seven angels that stood before God, who were given seven trumpets. The sound. of the harps. It has been heard on Earth. A-X-L-E, the x of gods ora KISS is l, levels e, eternal. Write down your life experiences, and wait until you send the correct one-to-one level. If the axle stops spinning, we need this way to keep safe and maintain the balance of weather as we go. Material possessions earth keeps here, and the unspoken words in time are spoken.

AXLE -A, NGELS. K, ISS X L,ORD E, ETERNITY

A REPETITION IN LIFE and our reassurance against each other over time. Then, I realised our words had more than one meaning. By breaking down the words into alphabet and numerology, we develop answers to the next level for human survival. And the passage through time. To be taken. Quote from the bible verse 10

The word= DICTIONARY=118=19=10=1=(19)psalms=The lord has punished me, but not handed me over to death. (10) Revelations = Then I saw an angel coming down from heaven surrounded by a cloud, with a rainbow over his head. His face shone like the sun. His feet flashed with fire. He set his right foot on the sea and his left foot on earth. He gave a great shout, like the roar of the lion. And the seven thunder crashed their reply. And I was about to write what the thunder shouted. When a voice from heaven called me, could you do that? The words are not to be revealed. She lifted her right hand to the heavens and swore by him, who lived forever and created heaven and everything in it. And earth and all their creatures and inhabitants. Then there should be no more delays when the seventh angel blew the trumpet—my stories through ages ever since his prophets will be fulfilled. This verse from the bible tells of the next level in time when we have to use our words to keep a balance out on the weather—Mother Nature; single letters.

DICTIONARY=.d, instance i, inner c, coarse i, n o, ver n, ext. A, ll- users y, yesteryear.

Or weather patterns - DICTIONARY d, own i,n c, entre t, turn I, n o,n north: a, ll, r, uler y, ears. A dictionary is a way to understand words and their meanings. But as our words, actions, and thoughts cling to our waist like a belt, we also have part bible verses clinging to our waist by breaking down each word that comes in good or bad, by numerology and alphabet. We come up with a number that represents a verse in the bible. Which lord god would you like us to use to balance out the weather (Mother Nature)? We have theories of what one mind's eye can see by vision or wonders. Men, women, one up, one down of understanding the passage on the level. This book gives others an idea of how to do it in a way that is fair to all cultures and humanity. As we go, we can balance the weather out with our words on the levels of all cultures. Patience is the key to everything we say and do in life.

PATIENCE.69-15-6=(69) PSALMS. The floods have risen. Deeper and deeper, I sink into the mire. The water rises around me. I wept until I was exhausted. (15) Don't let the floods overwhelm, or the ocean swallow me. (Many parents punish children angrily, which does not warrant it. If we lose patience, we make—mistakes. As we are not perfect, the past comes in on parents, too.

Revelations. Patience. (15) And I saw in heaven another mighty pageant showing things to come. Seven angels were assigned to carry the seven last plagues down to earth, and God's anger was finished. (6) As I watched, the lamb broke the first seal. Another theory is that the earth's seawater was around every planet nearby. But could it also go back around other planets? But they ended up down on earth. When we make a mistake and break the balance between each planet. we rotate our wise wisdom words to the next level once we release the tension, anger, and wars between Earth and the other planets. Would seas flow again around other planets nearby? This one must remember is just a theory of insight. Wonder and vision.

CHAPTER 2

MIND ONE'S MANNERS GIVE WITHOUT WANTING BACK: THE WAYS OF THE WORLD, OR MIND OVER MATTER.

ALL THESE WORDS TELL A story in verses of the bible. KNUCKLES = 94-13-4 plasms (94) lord god, to whom your vengeance belongs, Let Your glory shine. Arise, judge, the earth. (13) he helps us by pushing us to make us follow his path (4) hear their insolence and see their arrogance. How they boast, revelations - (13) and now in my vision, I saw a strange creature, rising from the sea with seven heads- ten horns ten crowns upon his horns, (4) there as I looked at a door standing open in heavens same voice sounded like a trumpet blast and said. COME! I will show you what must happen in the future. We need to use our words to balance the weather patterns. God outsmarted the devil this time, so all words, good and bad, can be used; when I say all words good or bad, this means once one checks out the verses that match the word in the bible, they can be used on levels of all cultures on the ground survival of human life and this planet. Our words are the next step in understanding the wise wisdom of every word, which matches a Bible verse. Tell precisely what one is looking for in answer. To improve on passage into our futures. The moral behind this book is to find an answer to life's most complex problems. Men have been looking for an answer for centuries, to no avail. But one way has been found: to the real reason for our being, how to proceed with time travel.; ve words to balance the weather in cloud forms of all earth; 's creation, forms images are in the cloud as images of words, action, thoughts, wonder, in-universe. Past words start to up to the reality of our lives. And why God made us in his image to help with his plan. Nursery rhymes, songs, and past sayings—vision, wonder— all play a part. When our forefather and mother started a sentence, they would forget the rest of it. A person would be interrupted while speaking. What they wanted to say, so we only got part of the story they tried to tell others so it could be improved later. There is an adult meaning to all nursery rhymes. To keep children›s minds down to levels they can understand, they lose concentration on their sentences once interrupted. And the thoughts about what would be said got lost in the interruption, and only half the sentence would get said. These half-sentences are called unspoken words, past and present. So, to unravel the whole picture and add the last sentence to the future, we turn and start again. We were old when we were born. We grow, learn, make mistakes, and learn from them. Pick up pieces on the way, I MAY HAVE REPEATED MYSELF IN THIS BOOK ON CERTAIN SENTENCES BUT ONE SEE IVE IMPROVE THE

SENTENCE FURTHER. Words have no wings but can fly for miles. My starter books now have my whole version of theories: To help others improve further on the grounds of the safety of this planet and all cultures. We once thought about it when I was a youngster. Why were some children white, and others had dark cultures? As I went to school with different cultures at that age, I could understand this, so it got put back in my memory bank until I was ready to find an answer. Well, I was forty years old when God sent me an answer. The answer came thick and fast. I do know how we became different cultures. Every chapter can tell a story about the representative verses in the Bible. Write down your thoughts and wonders about words from your childhood. Also, put the date when one first learned of theories and wonders. Vision words: as with all of us, only parts of the sentence come in. at any given time. Write down the time, date, and year of the acknowledgement. of any given situation idea. Words of forethought to further time. One person might know something early but can improve once it becomes known to another. Our words are the most essential part of this era of our lives. As we go, we balance out the weather. (Mother Nature) When Mother Nature helps to balance our words, it guides men back on the right path if they stray too far in the wrong direction before time—write some words down to see how they match the bible verse perfectly. Of what we determine the words mean. And how Bible verses send a clear view of real meaning. This book shows a way it can be done. In fairness to all cultures on levels first.

BUT IN THE FUTURE OR DEATH, IN OUR WORDS, THAT WILL CARRY WEIGHT FOR YEARS TO COME? God made us in his image to help him with his plans: conquer our words that form the clouds and the weather as we go.

Thoughtscape. =PSLAMS. (13) How long will you forget me? Lord, forever. How long will it take you to look the other way? When am I in need? How long must Ihide daily anguish in my heart? (4) Then, as I looked, I saw the door. Standing open in heaven, the same voice I heard before sounded like a trumpet. Blast. He spoke to me and said, Come up here; show you what must happen in the future. The gods got more intelligent than the devil, showing us how to use all words, good or bad, to balance out the weather patterns. So Noah's Ark days don't happen again. What is the mire, I might ask? It is where the devil lives, acts, and corrupts one's mind to commit wrongdoing against others or a crime. You might wonder what the word mire represents. This is how I came to understand how to use the bible verse to improve further about the real reason for our being. God never forgot anyone. Our mistakes, past and present, stood in the view of his messages, MIRE=13=9=18=5==4 5= Psalm righteous poem, as complete as words, a speechwriter pouring out his story, also the word AUDIOLOGY 108-18-9=revelations= (18) An angel came down from heaven to earth with great authority, and the earth grew bright with splendour. (9) Then the fifth angel blew his trumpet and saw one who had fallen to earth from heaven, and to her was given the key to the bottomless pit. To find the answer to the next step, balance out the weather patterns. As Winston Churchill Hill once said, a person who forgets his past has no future. We stopped learning and improving on the passage in time; God needed us to follow with our old-time songs, old sayings, and unspoken words and sentences that never finished because of an interruption while speaking. Entire sentences disappear from thoughts, and forgotten complete sentences never get repeated, becoming silent energy in the universe around us. This book is based on the day we were born. Our values are taught at home. All come from the bible verses. Then, when adults; words return to what we first knew, the Bible comes into play. The Bible can help the rest of the way. Our words, wisdom, and why must be used to help balance the weather (Mother Nature). The excellent book will help us understand the words spoken, good or bad, in all languages and how each verse tells us what the first meaning is in time, with words hanging over generations past and present. The reason behind this book and how it can be done is that we need to use all cultures to balance the weather patterns, our world action thoughts, theories, wonders, and most of all, patience. On levels. If any culture goes in any direction too soon, it brings stress on themselves; the more we learn, the more our bodies heal,

LIFE is for LIVING, GROW in MIND, BODY, and SOUL.

THE GOOD BOOK TELLS US how to understand the words spelt over centuries in different languages. The words change and get multiplied over the years, and other cultures are reused, which can be offensive to some. Words are not offensive if they are not directed at one another and used correctly. If they get directed at another person, like an insult, collect all the old-time sayings from your family tree first. Half of the sentence was never finished and can be completed by the coming generations. Jesus said, Love others as I love you. From the day we were born, we were born with a gift—the reason for our being. Our personalities, when used correctly on all levels, balance out the weather. as we go. These words mean to love others as I love you. What are the most profound meanings to them? How did Jesus mean it? How did he acquire this action? We kiss and cuddle other sweets, talk to them, or feel sorry for them, but is this the real meaning of Jesus' words of love? In my vision, I could see the key to how it could be done, in fairness to all cultures, and the meaning that could hold the key to our survival on earth. Of these words, Jesus spoke. I believe in the deep sense Of giving others what they need at any given time. We got so far in time that we lost our way. In God's words, he made us in his image to help with his plan: to conquer the whole universe as we go. At any given time, we provide another fundamental need: Balancing words to improve further in all cultures. Man has been looking for an answer for centuries, to no avail, but one thing has been found that will help us on the way to balancing out Mother Nature through verses in the bible. By matching an individual word, good or bad, I must say that all teaching children values in the home game are the verses in the bible. The black hole in the universe, Based on theory, God's words are in a verse in the bible. Return to what one first knew and then return to him (god). What he means by this is to go back to the beginning of one birth and back up the life one lived from birth to adulthood. Correct all mistakes, big or small, when they return at one. Correct the other person about what was said against or done to. Then, understand the wisdom of that mistake. One word in the verse of the bible can be explained by using the alphabet and numerology. As shown in this book then, checking the verse in the bible Is unbelievable how our words match up with the verse, explaining in detail a meaning that improves on what we knew thus far; it's a short, quick way to unravel the reason for our being . as always been stated. God made us in his image to help him with his plan. Each time we got thus far, The lord would have read the next step to improve further by balancing the weather on a level that included all cultures. It Will help save this planet and humans. And it could put sea levels around the nearest planets again. On the proper levels, each culture can keep the weather at bay. Still, we also must have a man-woman. Balance of level, One up and One down, and vice versa, without touching another level before one's time, by staying in one's family tree before going to the next level. Give our next generations something to live for, not die for. Let them feel they are part of improving our futures when we clean our bodies, not our minds. Material action and spiritual action: We leave our mistakes to others. But individuals can't open the brown door into heaven until all errors are corrected against others. In this book outlined how it can be done on a level for all cultures as well; we balance out the weather,

FIRE. SUN, WIND, RAIN, SNOW. THE FOUR SEASONS. The four angels at the corner of the earth help keep the balance F, S, W, R =Future sun wind ruler. Or words ruler-future sun. Take note of nursery rhymes, old-time sayings, songs, thoughts, and wonders. Vision and having one of them, as these theories could help improve further ahead in time. God doesn't punish us. We punish ourselves by going too far in the wrong direction before our time or levels. We punish ourselves when we don't stop thinking before we act. Or speak. The values we are taught in the home while growing up come from verses in the bible. Our words were used, in short, to guide us through life, spiritually and verbally.

CHAPTER 4

HOPE IS BEING ABLE TO SEE LIGHT IN ALL THIS DARKNESS

E ARE BORN OLD; WE grow in mind, body, and soul. Our mistakes as we grow up are the key to all these levels of understanding our words and their uses to help balance out clouds. Our forefathers and mothers, subconscious minds, knew part of the truth for us to follow. But our conscious minds got it wrong. How many times have we listened to others talk? How many times do we get interrupted before they say the whole sentence? This is how the catchphrase, or old-time saying, originated, as one starts saying something and the rest disappears from them. Where does the rest go? What do we all do as youngsters? We wonder. We think. Often, our vision will see things the mind doesn't understand. So, it gets put aside, and one thinks about returning to those thoughts another day. But as it happens, the day never comes. When we retire, we forget what we put out there that we wanted to know more about— our wonders, thoughts, and dreams. Our retirement is the best time to ponder past events and correct our mistakes against others. We don't need to look for answers or our mistakes; just wait, and they will surface; we might see someone do something similar. Or rethink our lives. If we correct past mistakes as they come into use in later years, we understand the wisdom of these mistakes in words. The old saying: Wherever one may be, let your wisdom go free. Says that wisdom is easy to carry but difficult to gather.How true this word past is, and how enlightening it is. For futures to improve further. The saying of words is easy to carry. They cling to the waist like a belt—a Lamb seal to tell the story. I began to unroll the scroll. The first was a white horse of daylight colour. The second was a red horse. The third was a black horse. 4th was a pale horse, which means death. 5th horse hell. Who was given control? One-quarter of the earth. The sun became dark in the sixth seal of vast earthquakes, and the moon became blood-red in the eclipse. We have just read these six horses, representing the colour. And different cultures are alive today. With respect, our words are so exciting and enlightening.

SHILL=S, SUNSHINE H, EVEN I, N L, ONG L, not just reading further understanding the depths of our words and how they can open up heaven gate to the universe.

BUT IN THE FUTURE OR DEATH, IN OUR WORDS, THAT WILL CARRY WEIGHT FOR YEARS TO COME? This phrase is accurate to the letter stating that we are doing wrong if we stop moving on God's path of survival. We die, but our bodies heal when we learn more and improve.

DEATH=38-11-2=(38) (PSALMS) Lord, you know how I long for my health once more; my heart beats wildly, and my strength fails, and I'm going blind. They call it retirement. Once in retirement, return to what you first knew, then come back up through your path, correcting all mistakes against other in-word deeds. Then, understand their wisdom. Then, the brown door will be opened into heaven's gates. The gods got more intelligent than the devil, showing us how to use all words, good or bad, to balance out the weather. Patterns. I have a granddaughter who has heard the trumpet out there, and this enlightenment encouraged me further into this vision and foresight into the future. I have 3listened to the trumpet sound, which tells me that what we older ones don't hear and see our next generation could improve our knowledge further of the reason for our being if we lay the groundwork for them to follow. Yes? What is the mire, one might ask? It is where the devil lives, acts, and corrupts one's mind to commit wrongdoing against others or a crime. You might wonder

MIRE=13=9=18=5= 5= Psalm righteous poem, as complete as words, a speedy writer pouring out his story, also the word (5) revelations =I saw the scroll in the right hand of the one sitting on the throne. The scroll had writing on the inside and the back, sealed with seven seals. Who is worthy to break this seal? Of the scroll and to unroll it. As no one anywhere was worthy, no one could tell us what it said. Was this what was in the scroll of this generation? How can we proceed in time, unravel, and use our words to balance weather as we go? And I wonder about the seven seals. The seven angels in the sky are stars, but one is never seen; this one was sent to Earth to help unravel the wisdom of our words. One must remember that this is written about vision, thoughts, wonder, and my own life experiences. Old-time sayings, nursery rhymes, songs,

CHAPTER 5

BRING THE WARMTH ENJOY WITHOUT COMPLAINT

Cleaning our bodies, not our minds, will not solve the world's weather problems. Any book that is written will always attract one's attention, like the verses in the Bible. There is always something in what's written in one's mind that one would want to know more about. They say the answers are all in the Bible. This is true when we break our words down. Then, could you match them with verses in numerology and the alphabet? No matter what one reads, there will always be something that enlightens one, whether it is the past or the present. An Asian friend mentioned that their culture was working on it to rain at night and fine during the day. This word inspired me to venture further into my theories, vision and wonder as a child—insight into what's out there. Also, when one speaks a sentence, it will always stretch further in a long conversation which gathers around our waist. This book explains how we can understand the meaning passage of our words and how the bible does have all the answers. Especially when it comes to balancing out the weather. Every word in all cultures forms clouds, creating the weather. In Jesus' words, love others as I love you. True love comes from the heart, not one's mouth. Give without wanting. Give what one needs at any given time. Not what one wants. Our unspoken words, like a rolling stone, gather no moss. It all ends up in cloud formations. The reason behind putting pen to paper in a book form is how we can define our words and understand the verses in the bible they match: alphabet numerology and technology. It will get us around the world fast with the balance of the weather as we go. as shown in this book of words.

We are born old and grow in mind, body, and soul. Creating our mistakes as we grow is the key to all these levels of understanding our word and its uses to help balance out clouds. Our forefathers and mothers, subconscious minds, know the truth, but our conscious mind gets it wrong. How many times have we listened to others talk? How many times have we been interrupted before One says the whole sentence? This is how the catchphrase, or old-time saying, originated: as one starts saying something, the rest disappears from them. Where does the rest go? I wonder what we all do as youngsters. We wonder, We Think. As often, our vision will see things the mind doesn't understand. So, what one thinks gets put aside, and one returns to the thoughts another day. But as it happens, the day never comes. When we retire, we forget what we put out there that we wanted to know more about—our wonders, thoughts, and dreams. Our retirement is the best time to ponder past events. In one's own life, correct all mistakes against others. We don't need to look for errors as they are all around our bodies. Just wait, and they will surface; we might see someone do something similar. Or rethink our lives. If we correct past mistakes as they come in using later years, we understand the wisdom of these mistakes as they return to our person, in words. The old saying (wherever one may be, let your wisdom go free) is = that wisdom is easy to carry but difficult to gather. How valid the words of the past are, and how enlightening.For futures to improve further. The saying of words is easy to carry and tell the story. They cling to the waist like a belt.

SHILL means in any language, bible verse? 60-6- (6) The lamb broke the scroll seal. I began to unroll the scroll. The 1st was a white horse. Daylight color. 2nd was a red horse. 3rd. was a black horse. 4th was a pale horse. Means death. 5th horse hell. Who was given control of one-quarter of the earth? The sixth seal opened vast earthquakes; the sun became dark, the moon's blood red .color. The eclipse. We just saw the 2024 June month. Meanings. These six horses all represent the colours of the universe. And different cultures are alive today. With respect, Our words are so exciting and enlightening.

SHILL=S, UN H, EAVEN I, N L, ONG L, IVES. Or it could be the seasons or input. S, south H, high I,n L,evel. L, eft. Not just reading the bible. It's understanding the verses of our word. The depths of our words further explain how they can be used to open up heaven's gate to the universe. The reason for our being, The white horse, is truth and honesty within ourselves. I have raised four children, two boys and two girls. I realised they all had different personalities. To handle. Each one handles problems in life differently than another. As my late husband drowned at an early age, I was left to raise them alone, without a father or guidance, with my parents, brothers and sisters. I learnt about the negative and positivity in nature, which is how I conquered their personalities and understood what was happening then. Every child born can follow one parent's wise words, or the other, in their side family wisdom of words, action, of the family tree. Some could handle both side's father and mother's wisdom from their family tree. Grow and learn further. I also put these phrases in this book, as it shows how the values we taught in the home all match up to verses in the bible. One way is to find and understand what each word means in God's strict path to follow and how to use them in our words.

Example= LEADERSHIP=97=16=7 (97) Let earth rejoice till the furthest island. To bring God clouds and darkness and surround God justice and foundation of his passage in time. (Psalms)revelations The mighty voice shouted to seven angels, now go out your way and empty seven flasks off the gods upon earth. In 2022, these plagues are surfacing in the public arena again. We may have stopped learning and improving further on the next level of our words to understand their wisdom. After all, once mistakes are corrected, as individuals against others, this is the key to future levels: Fixing our past and present mistakes. We open up further energy forces and spread wisdom around all cultures to improve further or to hold (patience). (7) I saw four angels standing at four corners of the earth, holding back four winds. This is called leadership with those who understand how the levels work. This is the level of rainbows in the skies that circle Earth between weather patterns. The values we taught are all in verses of the bible. Seven angels were given flasks to sprinkle diseases up the earth. Four angels were sent to the four corners of the earth for the four seasons to guide.

(1) Speak without accusing (James-1:19)

(2) Give without sparing or wanting (proverbs: 21: 26)

(3) Pray without ceasing with meanings. (Colossian's 1:19)

(4) Answer without arguing (proverbs (17:1)

(5) Share without pretending (Ephesians (4:15)

(6) Enjoy without complaint. (Philippians (2:14)

(7) Trust without waving. (Corinthians (13: 7)

(8) Forgive without punishing (13:12)

(9) Promise without forgetting (proverbs, (13 12)

(10) Listen without interrupting, (proverbs-18:7)

Don't stand behind me; I may not lead. Don't stand in front of me; I may not follow, but stand beside me and be my friend.

Water sign = cancer, Scorpio Pisces RAIN TEAR DROPS,

Earth sign, = Taurus, Virgo, Capricorn KEEP OUR FEET FIRMLY on the ground

Air sign = Libra Aquarius, Gemini. LET THE AIR THROUGH

Fire sign = Leo. Aries, Sagittarius, heat sunshine. Based on insight theory.

CHAPTER 6

ANSWER WITHOUT ARGUING

E ARE BORN OLD. WE grow in mind, body, and soul; our mistakes as we grow up are the key to all these levels of understanding of our word and their uses to help balance clouds. Our forefathers and mothers, subconscious minds, knew the truth, but our conscious minds got it halfway wrong. How many times have we listened to others talk? How many times have we been interrupted before they say the whole sentence? This is how the catchphrase, or old-time saying, originated: as one starts saying something, the rest disappears from them. Where does the rest go? I wonder as we all do as youngsters. We wonder, We Think. Often, our vision will see things the mind doesn't understand. So, it gets put aside; one thinks and returns to those thoughts another day. But as it happens, the day never comes. When we retire, we forget what we put out there we wanted to know more about—our wonders, thoughts, and dreams. Our retirement is the best time to ponder past events. Of one's own life, correct all mistakes against others. We don't need to look for our mistakes as they are all around our bodies. As good books say, they cling to our waist like a belt. Just wait, and they will resurface; we might see someone do something similar. Or rethink our lives. If we correct past mistakes as they come in using later years, we understand the wisdom of these mistakes as they return to our person, in words. The old saying (wherever one may be, let your wisdom go free) is = that wisdom is easy to carry but difficult to gather. How valid the words of the past are, and how enlightening. For futures to improve further. The saying of words is easy to carry and tell the story. They cling to the waist like a belt. Horse. Daylight color. 2nd was a red horse. 3rd. was a black horse. the 4th was a pale horse. Means death. 5th horse hell. Who was given control of one-quarter of the earth? The sixth seal opened vast earthquakes; the sun became dark, the moon blood red. Colour. The eclipse. We just saw the 2024 June month. Meanings. These six horses represent the colours of the universe. And different cultures are alive today. With respect, Our words are so exciting and enlightening.

L, ONG L, IVES. It's not just reading the bible. Further understanding the depths of our words and how they can open heaven's gate to the universe and other planets. The good and evil in us all. But we must do it on a level that all culture's wise words follow. To bring the horoscope into this learning process, we all have a negative side and a positive side; once we understand the positive side and separate the negative from our thoughts and actions, we all have a great personality created to handle every situation. This personality and the opposite of our signs create a whole nature—the reasoning behind why we should wait for the right eternal partner so as past mistakes. The past can't get through to upset our feelings thus far. In part of the horoscope, I have laid out signs opposite to what I've learned thus far. Each one has a role in our lives and is the energy force of the sky. Fire signs bring heat and sun warmth, and Water signs bring rainstorm drizzle. Earth signs keep the evenness offset firm on the ground. Air Signs bring in fresh, cold air around people. We would only move when it was our level, too. With opposites above us, we do not go to extremes in any one direction; relationships are built on trust.

THE 7 STARS, 7 GODS, 7 ANGELS, 7 PLANET 4 -7,s =28-10-1(10)Now we saw another angel coming down from heaven, surrounded by a cloud. With a rainbow over his head, his face shone like the sun, and his feet flashed with fire. He held in his hand a small scroll; he set his right foot on the sea, his left foot on the earth. And the seven thunder crashed their reply. Was this small scroll carrying the message to use our words to balance out the weather (mother nature) 3 seven 21 the key to understanding the real reason for our being? We lived before, made a mistake, and broke the universe apart four times; each time, we had to start from scratch again. Hopefully, people will notice how meaningful our lives are this time and help balance out weather patterns, Mother Nature, and the level of one family tree first. Then, our words will be used to advance weather patterns.

CHAPTER 7

THERE IS A LIGHT IN ALL THIS DARKNESS SPEAK WITHOUT ACCUSING

GOD HAS SEVEN STARS IN the sky, but one has never been seen from Earth; I believe it is the star standing beside god to spread his words of wisdom ahead of us. Be sent down to the pits of hell to find help, and others can follow the right path. To level others can understand, once all past mistakes are corrected back to the person done to or family. Lots of us sacrificed material possessions to understand the real reason for our being. Man has been looking for a way to stay alive in human form, for we stopped earning and improving with time; God needed us to follow with our old-time songs, old sayings, and unspoken words and sentences that never finished because of an interruption while speaking. Entire sentences disappear from thoughts, and forgotten complete sentences never get repeated, becoming silent energy forces in the universe around us. This book is based on the day we were born. Our values are taught at home. All come from the bible verses. Then, when adult words return to what we first knew, the Bible comes into play. The Bible can help the rest of the way. Our words, wisdom, and why must be used to help balance the weather (Mother Nature). The excellent book will help us understand the words spoken, good or bad, in Languages and how each verse tells us what the first meaning is in time, with words changing over generations past and present. The reason behind this book and how it can be done is that we need to use all cultures to balance the weather patterns, our world action thoughts, theories, wonders, and most of all, patience. On levels. If any one culture goes in any direction too soon, it brings stress on themselves; the more we learn, the more our bodies heal,

LIFE is for LIVING, GROW in MIND, BODY, and SOUL.

The good book tells us how to understand the words spelt over centuries in different languages. The words change and get multiplied over the years, and other cultures are reused, which can be offensive to some. Words are not offensive if they are not directed at one another and used correctly. If they get directed at another person, like an insult, collect all the old-time sayings from your family tree first. They are half sentences that were never finished and can be finished by the coming generations. Jesus said, Love others as I love you., From the day we were born, we were born with a gift—the reason for our being. Our personalities, when used correctly on all levels, balance out the weather. as we go. These words mean to love others as I love you. What are their most profound meanings? How did Jesus mean it? How did he acquire this action? We kiss and cuddle each other and sweets, talk to them, or feel sorry for them, but what is the real meaning of Jesus' words of love? In my vision, I could see the key to how it could be done, in fairness to all cultures, and the meaning that could hold the key to our survival on earth. Of these words, Jesus spoke. I believe the deep sense is To give others what they need at any given time. We got so far in time that we lost our way. In other words, he made us in his image to help with his plan: to conquer the universe as we go. At any given time, we provide another fundamental need: balancing words to improve further in all cultures. Man has been looking for an answer for centuries, to no avail, but one thing has been found that will help us on the way to balancing out Mother Nature through verses in the bible. By matching an individual word, good or bad, all teaching children's values in the home match those in the bible.is a verse in the bible. The black hole in the universe, Based on theory, God's words are in Bible verses. Return to what one first knew and then return to him (god). What he means by this is to go back to the beginning of one birth and back up the life one lived from birth to adulthood. Correct all mistakes, big or small, when they return at one. Correct, go back to the one it was said or done. Then, understand the wisdom of that mistake. One word in the verse of the bible can be explained a lot by using the alphabet and numerology. As shown in this book, checking the Bible verses is unbelievable, especially how our words match up with the individuals. Verses explain in detail a meaning that improves on what we know thus far; it is a short, quick way to unravel the reason for our being, as always stated. God made us in his image to help him with his plan. Each time we got thus far, The lord would have ready the next step to improve further by balancing the weather on a level that included all cultures. It Will help save this planet and humans. And it could put sea levels around the nearest planets again. On proper levels, each culture can keep the weather at bay. Still, we also must have a man-woman. Balance of level, One up and One down, or vice versa, without touching another level before one time, by staying in one family tree before going to the next level. Give our next generations something to live for, not die for. Let them feel they are part of improving our futures when we clean our bodies, not our minds. Material action and spiritual action: We leave our mistakes to others. However, any individual can open the brown door into heaven once almost all past mistakes in words and actions are corrected until all errors are corrected against others. This book outlined how it can be done on a level for all cultures as we go: we balance out the weather.

FIRE. SUN, WIND, RAIN, SNOW. THE FOUR SEASONS. The four angels at the corner of the earth help keep the balance F, S, W, R =F, future s,un w, and r, ruler. Or words ruler-future sun. Take note of nursery rhymes, old-time sayings, songs, thoughts, and wonders. Vision and having one theory on them, as these theories could help improve further ahead in time. BUT IN THE FUTURE OR DEATH, IN OUR WORDS, THAT WILL CARRY WEIGHT FOR YEARS TO COME? These are the vital sentences we never hear spoken very often and must heed in this life if we want to survive on earth in human form.

This phrase is accurate to the letter stating that we are doing wrong if we stop improving on God's path of survival. We die, but our bodies heal when we learn more and improve. DEATH=38-11-2=(38) (PSALMS) Lord, you know how I long for my health once more; my heart beats wildly, my strength fails, and I'm going blind. They call this retirement. Once in retirement, return to what you first knew, then come back up through your path, correcting all mistakes against other in-word deeds. Then, understand their wisdom. Then, the brown door will be opened into heaven's gates.

The gods got more intelligent than the devil, showing us how to use all words, good or bad, to balance out the weather. Patterns I have a granddaughter who has heard the trumpet out there, and this enlightenment encouraged me to look further into this vision foresight into the future; what we older ones do hear see our next generation could improve our knowledge further of the reason four our being if we lay the groundwork for them to follow. Yes? What is the mire, one might ask? It is the devil's life, words, and acts, corrupting one's mind to commit wrongdoing against others or a crime. You might wonder what the word mire represents in wisdom words.

MIRE=13=9=18=5= 5= Psalm a righteous poem, as complete as words, a speedy writer pouring out his story, also the word (5) revelations =I saw the scroll in the right hand of the one sitting on the throne. The scroll had writing on the inside and the back, sealed with seven seals. Who is worthy to break this seal? Of the scroll and to unroll it. As no one anywhere was worthy, no one could tell us what it said. Was this in the scroll of this generation? How can we proceed further in time, unravel, and use our words to balance weather as we go? And I wonder about the seven seals. The seven angels in the sky are stars, but one is never seen; this one was sent to Earth to help unravel the wisdom of our words. One must remember that this is written about vision, thoughts, wonder, and my own life experiences. Old-time sayings, nursery rhymes, songs,

CHAPTER 8

ENJOY WITHOUT COMPLAINT; BRING THE WARMTH

E ARE BORN OLD AND grow in mind, body, and soul. Our mistakes as we grow up are the key to all these levels of understanding our word and their uses to help balance out clouds. Our forefathers and mothers, subconscious minds, know the truth, but our conscious mind gets it wrong. How many times have we listened to others talk? How many times have we been interrupted before they say the whole sentence? This is how the catchphrase, or old-time saying, originated: as one starts saying something, the rest disappears from them. Where does the rest go? I wonder as we all do as youngsters. We wonder, We Think. As often, our vision will see things the mind doesn't understand, what one thinks gets put aside, and one returns to the thoughts another day. But as it happens, the day never comes. When we retire, we forget what we put out there that we wanted to know more about—our wonders, thoughts, and dreams. Our retirement is the best time to ponder past events. Of one's own life, correct all mistakes against others. We don't need to look for errors as they are all there around our bodies. Just wait, and they will surface; we might see someone do something similar. Or rethink back on our lives. If we correct past mistakes as they come in using later years, we understand the wisdom of these mistakes as they return to our person, in words. The old saying (wherever one may be, let your wisdom go free) is = that wisdom is easy to carry but difficult to gather. How valid the words of the past are, and how enlightening.For futures to improve further. The saying of words is easy to carry and tell the story. They cling to the waist like a belt. May I ask what the word = SHILL means in any language, bible verse? 60-6- (6) The lamb broke the scroll seal. I began to unroll the scroll. The 1st was a white horse. Daylight color. 2nd was a red horse. 3rd. was a black horse. 4th was a pale horse. Means death. 5th horse hell. Who was given control of one-quarter of the earth? The sixth seal opened vast earthquakes; the sun became dark, the moon's blood red .color. The eclipse. We just saw the 2024 June month. Meanings. These six horses all represent the colours of the universe. And different cultures are alive today. With respect, Our words are so exciting and enlightening.

SHILL=S, UN H, EAVEN I, N L, ONG L, IVES. Or it could be the seasons or input. S, south H, high I,n L,evel. L, eft. It is not just reading the bible. It understands the depths of our words further and how they can be used to open up heaven's gate to the universe. The reason for our being, The white horse, is truth and honesty within ourselves. I have raised four children, two boys and two girls. I realised they all had different personalities. To handle. Each one handles problems in life differently than another. As my late husband drowned at an early age, I was left to raise them alone, without a father's guidance, with my parents and brothers and sisters. I learnt about negative positivity in nature, which is how I conquered their personalities and understood what was happening at the time with them. Every child born can follow one parents's wise words, or the other, in their side family wisdom of words, action, of the family tree. Some could handle both sides father and other wisdom from their family tree. Grow and learn further. I also put these phrases in this book, as it shows how the values we taught in the home all match up to verses in the bible. One way is to find and understand what each word means in God's strict path to follow and how to use them in our words.

It is called repetition. It is exciting and enlightening the reason for our being. So, let's try these words: PARROTS. Researched, and some live till 80-odd years old. We are now having a parrot in the home. It repeats itself with our words, and humans make the same mistakes from the past without knowing what or why they do—called repetition in life. PARROTS= (17) Other fools are ill because of sinful ways; their appetite is gone, and death is near. They cried, lord, in their troubled time, help and deliver them. Spoke, they healed, snatched from the door of death; my theory is this verse explains what happens if we stop learning about the real reason for our being when we retire, we give up words, keep changing, and elderly ones in retirement forget the wise words they sowed. Ill health sets in, and the mind gives up. So, by writing down whatever comes to a head, good or bad words break down into numerology and whatever verse in any chapter or verse in the bible, the bible does have all our answers. In individual verses. Of words. If one takes the word, LIE=26-8) dismiss all charges against me, for I have tried to keep your laws and trusted god without waving. (8) For I have washed my hand to prove my innocence., As I see it, when we try to understand the real reason for our being, we try to understand everything at once, but it is best only to consider what one needs to know at any given time. God won't send one answer to one wonder or thought from birth until the time comes when one's mind is ready to understand and improve one's father. The white horse is truth and honesty within ourselves. I have raised four children, two boys and two girls. I realised they all had different personalities. To handle. Each one handles problems in life differently than another. As my late husband drowned at an early age, I was left to raise them alone, without a father's guidance, with my parents and brothers and sisters. I learnt about negative positiveness in one's nature; this is how I conquered their personalities and understood what was happening then. Every child born can follow one parent's wise words, or the other, in their side family wisdom of words, action, of the family tree. Some could handle both sides, father and mother wisdom from their family tree. Grow and learn further. I also put these phrases in this book, as it shows how the values taught in the home all match up to verses in the bible. One way later to find and understand what each word means in God's true path to follow—our words. Wisdom.

LEADERSHIP=97=16=7 (97) Let the earth rejoice to the furthest island.to bring Clouds and darkness and surround God with justice and the foundation of his passage in time. (Psalms) Revelations=verse(16) The mighty voice shouted to seven angels, now go out your way and empty seven flasks off the gods upon earth. In

2022, these plagues are resurfacing in the public arena again. We may have stopped learning and improving further on the next level of our words to understand their wisdom. After all, once mistakes are corrected, as individuals against others, this is the key to future levels: fixing our past and present mistakes. We open up further energy forces and spread wise wisdom around all cultures to be improved further for the next generations. Or to hold (patience). (7) I saw four angels standing at four corners of the earth, holding back four winds. This is called leadership with those who understand how the levels work. This is the level of rainbows in the skies that circle Earth between weather patterns. The values we taught are all in verses of the excellent Bible book

CHAPTER 9

THE WATERWAYS

Walking through water Enjoy without complaint

*O*F THE SPINNING PLANETS.MY FORESIGHT of how it might have been.

THE SONG= RAINDROPS KEEP FALLING ON MY HEAD, or was it word drops? Raindrops and teardrops are the most common things in our daily lives today. Still, in reality, if words were raindrops, they could have meant WORD DROPS ON MY HEAD, and how we sent our wise words over to the next level gets improved on and back again. Words on ruler distance

WORDDROPS.=132=16=7=(132-)lord, remember when my heart was in turmoil; I couldn't rest, sleep, or think there had to be more to our lives. And a way to use our words, good or bad, to keep the balance of weather patterns. In the past, the ark was built to protect it from extreme weather. Now, in this day and age, another way is to use our words, good or bad, to balance the weather in all cultures worldwide. Will it be possible to spread the rising seas on Earth around another planet as we use up the words wisdom tension between us and the other planets? Mars, Venus, Mercury, and Earth only have life on them. If the seas had been around these planets before we broke the universe apart, seawater would have blocked the flow of seawater only around Earth. This is a theory worth considering as if my theory is correct and the ring around other planets is its weather from pasts. By unravelling our wise words, we could put weather back around other planets as it was.= And then heard a mighty voice. We were shouting from the temple to the seven angels. Now go your ways and empty seven flasks of the wrath of god upon the earth. The first angel poured her flask over the planet, and sores broke out.2nd angel poured over the ocean; it became like water and blood, and all in the sea died. 3rd angel upon rivers springs you are just in sending this, judgement, the fourth angel upon the sun causing it to scorch all with its fire. The fifth angel kingdom plagues into darkness, and the sixth angel, the great river, dries up. Seventh angel into the air, it is finished. These words from the Bible created our four seasons: thunder, lightning, and extreme weather. So, using our words today to balance the extreme weather will outsmart the devil's corruption against others. As scientists discover more about what is out in outer space, there are more planets on which we can live. But would it be logical to rotate the weather around them first? Before they become livable. If it is possible to stretch out between each, our words are around each planet in cloud rings—and tension forces. We rotated and used our words to take tension away, and the sea level spilt around other planets, as in past times. The possibility is that we have lived on these planets before. This is based on theory -insight and vision of past and present words. that create the clouds. If the earth is flat, this solves many unanswered questions about rising water and when and if the planet stops spinning on its axle. What then? Do we give up? No, we found a way to create a balance towards our close planets. Mars, Mercury, Venus. Earth. by unravelling our past and present words and understanding their wisdom after we correct all past mistakes against others. We can balance the weather (called mother nature). Once we leave the tension words in cloud forms, we stay on a level with all cultures. Family tree first. Then, up one step time by lifting our word around these other planets and letting the seas flow around them, like at the beginning time, by lifting the tension words and images further out, we will be able to let the rising seas on Earth flow through, around create growing plant trees, hoping the weather above will make the rain, wind, sun, as on earth. Every idea, theory, insight, wonder, and vision will play a part in unravelling the real reason for our being. Our words and action deeds all form images in the clouds. Raindrops and teardrops are The most common thing that happens in our daily lives today, But in reality, Were words the raindrops, they could have meant WORD DROPS ON MY HEAD and how we sent our wise words over to the next level get improved and back again—words on ruler distance sentences.

WORDDROPS.=132=16=7=(132-) lord, remember when my heart was in turmoil; I couldn't rest, sleep, or think. There had to be more to our lives. And a way to use our words, good or bad, to keep the balance of weather patterns. In the past, the ark was built to protect it from extreme weather. Now, in this day and age, another way is to use our words, good or bad, to balance the weather in all cultures worldwide. Will it be possible to spread the rising seas on Earth around another planet as we use up the words wisdom tension between us and the other planets? Mars, Venus, Mercury and Earth.

I would only have lived it. If the seas were around these planets before we broke the universe apart, The blocked sea water from flowing only around Earth and other planets. This is a theory worth considering as if my theory is correct and the ring around other planets is half done from the past. Then, by unravelling our wise words, we could put weather back around other planets as before revelations(16)And then heard a mighty voice. We were shouting from the temple to the seven angels. Now go your ways and empty seven flasks of the wrath of god upon the earth. The first angel poured her flask over the planet, and sores broke out.2nd angel poured over the ocean; it became like water and blood, and all in the sea died. 3rd angel upon rivers springs you are just in sending this, judgement, the fourth angel upon the sun causing it to scorch all with its fire. The fifth angel kingdom was plagued with darkness, and the sixth angel river dried up. Seventh angel into the air, it is finished—these words from the Bible created our four seasons: thunder, lightning, and extreme weather. So, using our words today to balance the extreme weather will outsmart the devil's corruption against others. As scientists discover more about what is out in outer space, there are more planets on which we can live. But wouldn't it be logical to rotate the weather around them first? Before they become livable. If it is possible to stretch out between each plant, the energy force that keeps the seawater down on one planet, if our words are around each planet in clouds rings. And tension forces. We rotated and used our words to take tension away, and the sea level spilt all around other planets, as it had been in past times. The possibility we lived on these planets before is based on theory -insight and vision of past and present words, that create the clouds. If the earth is flat, this solves many unanswered questions about rising water and when and if the planet stops spinning on its axle. What then. Do we give up? No, we found a way to create a balance towards our closest. Planets. Mars, Mercury, Venus. Earth. by unravelling our past and present words and understanding their wisdom after we correct all past mistakes against others. We can balance the weather (called mother nature). Once we leave the tension words in cloud forms, we stay on a level with all cultures. Family tree first. Then, up one step time by lifting our word around these other planets and letting the seas flow around them, like at the beginning time, by lifting the tension words and images further out, we will be able to let the rising seas on Earth flow through, around create growing plant trees, hoping the weather above will make the rain, wind, sun, as on earth. Every idea, theory, insight, wonder, and vision will play a part in unravelling the real reason for our being. Our words and action deeds all form images in the clouds.

CHAPTER 10

A FAR SIGHTED THEORIES IN MY MIND EYES

HAT IS THE PURPOSE OF using the bible verses to unravel the universe and everyday education? We all get parts of the bible coming in on us occasionally. And this verse, our minds understand . can cause tension, accidents, and retaliations against others, The bible verses. It would be easier for you to understand. Than understanding different people's far-sighted theories. Could it be, or have we had it all before? We made a mistake and broke the planets apart thus far. My theory of wonder and thoughts vision is once we collect all our words, past and present, around all cultures. For a level understanding of wise wisdom words, check out the verse in the bible by using the alphabet and numerology. Break each number down, and the Bible checks verses that suit one's family tree or wise word understanding levels. If we understand our words and have corrected our mistakes against others, then understand their wisdom on all cultural levels, we could open up passages around our nearest planets. Mar Venus mercury up to Jupiter Has earth accumulated all the water as the tension of our words, and anger, stress wars. Over centuries. All in cloud form, some images. Can the seas flow again around these planets if we part all these words, past and present? Like Jesus did when he raised the sea to allow people through. But would we reach a standstill in the universe? So, does our movement have to be on levels to maintain balance? When and if the axle stops spinning. Humans could rotate the words around on level all cultures to balance the sea weather.

WORDS=W-23=O15=R18 D-4=S-19=79-16-7(79) PSALMS: Your land has been conquered. In this heathen nation, our lives have been defiled and are in heaps of ruin.

=W, ISDOM O, N R, ULERS D, ISTANCE. S, UN. We want to show people how it can be done in all cultures. As we go, we keep a balance on weather patterns. If we had it all before, the answers are out there, waiting to be unravelled again. Do our words and reason cause life and death, and how can it be done? In fairness to all people and cultures. I've always wondered if there was anything more to our lives. What we believe is living. Little did I know that the Bible verses could be used to understand the real reason for our being. Looking at the whole picture ahead and behind it, I see that my path in time took a turn in a specific direction. I had planned for my future. I wanted to save, travel, or maybe go into politics, as I grew up having political conversations in the house. But we had to find the answer to life's most complex problems somewhere along the line. But as a teenager, like most people my age, I went out and about with friends. One night, my life took a turn. I became a rape victim, which changed the course of my plans. Immensely, I was married at 16 for eight years, and the first child was born in 1962. Three more followed, and I lost my late husband in 1970. While raising four children alone, I began to start thinking about why God created us and what the reason God created us was. I see friends and family all running around and doing the same things against each other, getting nowhere, and repeating the same mistakes against others. As my wonder grew, I realised there had to be more to our lives than what we were experiencing. I heard a lot. We grew up hearing these wise old sayings. What was the reasoning behind them? That being said, there just had to be a purpose. With a lot of trial and error over the years and putting it all together, I've found the way. The repetition of words in life will happen over the next generation. The same old things occur time after time and again in a generation if not corrected. The message of wisdom and unspoken words came to me. I put the pieces together in my family tree and my mind. The more I mused, the more I thought there had to be a way. As I said before, I probably repeat sentences in chapters, but that's okay for each time I improve or add more to the sentence. The moral behind this book is to find an answer to life's most complex problems. A man has been looking for an answer for centuries, to no avail. But one way has been found: to the real reason for our being, how to proceed with time travel. A way has been found to help us balance Mother Nature. The devil got outsmarted by god, and ideas to use our words to balance the weather cloud forms, all earth creation, forms images in the cloud images. Of words, action thoughts wonder. in universe. Past words start to add up to the reality of our lives. And why God made us in his image to help with his plan; nursery rhymes, songs, and past sayings—vision, wonder—all play a part. When our forefather and mother started a sentence, it was interrupted, and then they would forget to rest and say what they wanted to say, so we only got part of the story they wanted to tell others. There is an adult meaning to nursery rhymes. To keep children's minds down to levels they can understand, one loses concentration on their sentences once interrupted. And the thoughts about what would be said got lost in the interruption, and only half the sentence would get said. These half-sentences are called unspoken words, past and present. So, to unravel the whole picture and add the last sentence to the future, we return and start again. We were old when we were born. We grow, learn, make mistakes, and learn from them. Pick up pieces on the way back up through our lives to age retirement. I may have repeated myself in this book, but this is normal, as some sentences could come back that need improving further. I have repeated some chapters from my first two books. LIFE MOST COMPLEX PROBLEMS, AUTHOR BETSYBOO, and LIFE BIGGEST JIGSAW PUZZLE. AUTHOR Elizabeth Cargill Campbell. as each one leads into the next for improvement. My starter books now have my full vision of theories: to Help others improve further on the grounds of the safety of this planet and all artificial cultures. One thought when I was a youngster. Why were some children white, and others had dark cultures? As I went to school with different cultures at that age, I couldn't understand this, so it got put back in my memory bank until I was ready to find an answer. Well, I was forty years old when God sent me an answer. The answer came Thick and fast. Please accept my beliefs. I do know how we became different cultures.Every chapter can tell a story about the representative verses in the Bible. Write down one thought and wonder about words that come back. Also, put the date when one first learned of theories and wonders. Vision words: as with all of us, only parts of the sentence come in. at any given time. Write down the time, date, and year of the acknowledgement. of any given situation idea. Words of forethought to further time. One person might know something early but can improve once it becomes known to another. Our words are the most essential part of this era of our path if we stray too far in the wrong direction.

CHAPTER 11

SHARE WITHOUT PRETENDING HOW IT CAN BE DONE.

BUT IN THE FUTURE OR DEATH, IN OUR WORDS, THAT WILL CARRY WEIGHT FOR YEARS TO COME? How accurate are these words REASON FOR OUR PAST MISTAKES IN WORDS

THOUGHTSCAPE. =PSLAMS. (13) How long will you forget me? Lord, forever. How long will it take you to look the other way? When am I in need? How long must I hide daily anguish in my heart? (4) Then, as I looked, I saw the door. Standing open in heaven, the same voice I heard before sounded like a trumpet Blast. He spoke to me and said, Come up here; show you what must happen in the future. The gods got more intelligent than the devil, showing us how to use all words, good or bad, to balance out the weather. Patterns? What is the mire, one might ask? Where the devil lives, acts, and corrupts one's mind to commit wrongdoing against others or a crime. You might wonder what the word mire represents in wisdom words. MIRE=13=9=18=5==4 5= Psalma righteous poem, as complete as words, a speedy writer pouring out his story, also the future

We stopped learning and improving on the passage in time; God needed us to follow with our old-time songs, old sayings, and unspoken words and sentences that never got finished because of an interruption while speaking. Complete sentences disappear from thoughts, and forgotten entire sentences never get repeated, becoming silent energy in the universe around us. But to no avail. Only to have what we have done thus far wasn't good enough to keep human life alive; my theory on this is, and what I've seen by the vision, in past times, we made a mistake, and the elements of the universe suffocated earth, we had to start from scratch again four times in all. (About) this is one way it can be done. This is only the tip of the iceberg when opening up heaven and earth to our words, deeds, and actions. The black holes that they compounded were the words anger tension from cultures that shunned away from the family lifestyle. May the nursery rhyme, Merry Go Round the Moulder Bush, arise from these actions past once the black hole came about and blocked everyone outside. We should never eliminate material possessions, labour, and such; if the worst ever happens, we always find our way back. Our words are the most crucial passage over time. We seem to forget that the almighty god made us in his image to help him with his plan to conquer the whole universe (REMEMBER), As we have had it all before. I made a mistake and broke the energy force between each planet apart. And we had to start from scratch again. Christianity is based on the values we were taught as children; we had to live by them; the gods say to return to what one first knew and start again. Hemeans birth; as individuals, we are old, come back up through our life span and correct all mistakes against others back to the person they did against. Some of us sacrifice material possessions to understand the real reason for our being. Man has been looking for a way to stay alive in human form for centuries but to no avail. Only to have what we have done thus far wasn't good enough to keep human life alive. My theory on this is that what I've seen by the vision is that in past times, we made a mistake, and the elements of the universe suffocated Earth; we had to start from scratch again four times. (About)This is one way the walking stitch takes one step toward improvement. On the wild side, it depends on one walking stick, not others. The white horse is the truth, the light, and the way; first, be honest with yourself. Live by values taught in the home. Use patience in everything one does. Stop, look and listen to the same step we take to cross the road. Material possession, and other, is the spiritual way, unspoken words. If they say they found another planet like Earth out of the universe, there would be no life or trees, if this is true. How did it get there, and why are questions to be answered here? By having theories and research, we can discover the reason. Is it just an image or reflection of the earth before humans got their hands on it? Or does it exist? I might add a couple of theories: we lived on this planet before, and by going too far too soon in the wrong direction, we took the light out of the day, and the earth got stuck into the black, dark universe. As will happen in today's life, we go too far off our levels and go in the wrong direction. The other reason could be a planet in waiting, as under the word,

FLAGS (means all culture) (46=10=1) verse God is our help in times of trouble, and we need not fear if the world blows up. Could scientists see this planet as a stepping stone over when humans destroy each other? This word matches the bible verses and what they could mean for our future travels in the time of learning. So, we can put their pieces together by having theories on what our vision sees or wonders about. There always has to be work to be done. Never take life for granted, as we don't know what's around the corner. But by keeping an open mind on the world's ways, we can be prepared for the unexpected; as we are taught as children, god has a big book in the sky universe of all we think and say. We can only see as far as the mistakes we make on others or ones that cling to our waist like a belt. Understanding the way to use our words, good or bad, comes to us from our father and mother's parents, parents, brothers, and sisters. When we break words down, check the Bible verse to give one a clear vision of what words mean. I first stumbled upon this idea to get a better meaning to reason for our lives back when given an updated English bible. Another bible might have different verses with the same meaning but just worded differently—older time.

CHAPTER 12

TRUST WITHOUT WAVING.

*I*T IS CALLED REPETITION. IT is exciting and enlightening the reason for our being. So, let's try these words: PARROTS. Researched and would live till 80-odd years old. We are now having a parrot in the home. It repeats itself with our words, and humans make the same mistakes from the past without knowing what or why they do—called repetition in life.

PARROTS= (17) Other fools are ill because of sinful ways; their appetite is gone, and death is near. They cried, lord, in their troubled time, help and deliver them. Spoke, they healed, snatched from the door of death; my theory is this verse explains what happens if we stop learning about the real reason for our being when we retire, we give up words, keep changing, and elderly ones in retirement forget the wise words they sowed. Ill health sets in, and the mind gives up. So, by writing down whatever comes to a head, good or bad words break down into numerology and whatever verse in any chapter or verse in the bible, the bible does have all our answers. In individual verses. Of words. If one takes the word,

LIE=26-8) dismiss all charges against me, for I have tried to keep your laws and trusted god without waving. (8) For I have washed my hands to prove my innocence., As I see it, when we try to understand the real reason for our being, we try to understand everything at once, but it is best only to consider what one needs to know at any given time. God won't send one answer to one wonder or thought from birth il the time comes when one's mind is ready to understand and improve one's father. The white horse is truth and honesty within ourselves. I have raised four children, two boys and two girls. I realised they all had different personalities. To handle. Each one handles problems in life differently than another. As my late husband drowned at an early age, I was left to raise them alone, without a father's guidance, with my parents and brothers and sisters. I learnt about negative positiveness in one's nature; this is how I conquered their personalities and understood what was happening then. Every child born can follow one parent's wise words, or the other, in their side family wisdom of words, action, of the family tree. Some could handle both sides, father and mother's wisdom from their family tree. Grow and learn further. I also put these phrases in this book, as it shows how the values taught in the home all match up to verses in the bible. One way later to find and understand what each word means in God's true path to follow—our words. Wisdom.

LEADERSHIP=97=16=7 (97) Let the earth rejoice till the furthest island.to bring Godclouds and darkness and surround God's justice and foundation of his passage in time. (Psalms) Revelations=verse(16) The mighty voice shouted to seven angels, now go out your way and empty seven flasks off the gods upon earth. In 2022, these plagues are resurfacing in the public arena again. We may have stopped learning and improving further on the next level of our words to understand their wisdom. After all, once mistakes are corrected, as individuals against others, this is the key to future levels: fixing our past and present mistakes. We open up further energy forces and spread wise wisdom around all cultures to be improved further for the next generations. Or to hold (patience). (7) I saw four angels standing at four corners of the earth, holding back four winds. This is called leadership with those who understand how the levels work. This is the level of rainbows in the skies that circle Earth between weather patterns. The values we taught while growing up are all in the excellent Bible book verses.

CHAPTER 13

LISTEN WITHOUT INTERRUPTING

ORDS, WISDOM, THE TENSION BETWEEN us and the other planets? Mars, Venus, Mercury, and Earth could have had life on them. If the seas were around these planets before we broke the universe apart, seawater would have blocked the flow of seawater only around Earth. This is a theory worth looking at or thinking about as if my theory is correct and the ring around other planets is it weather did, from pasts., the by unravelling our wise words we could put weather back around other planets as was.= revelations(16)And then heard a mighty voice. We were shouting from the temple to the seven angels. Now go your ways and empty seven flasks of the wrath of god upon the earth. The first angel poured her flask over the planet, and sores broke out.2nd angel poured over the ocean; it became like water and blood, and all in the sea died. 3rd angel upon rivers springs you are just in sending this, judgement, the fourth angel upon the sun causing it to scorch all with its fire. The fifth angel kingdom plagues into darkness, and the sixth angel's excellent river dries up. Seventh angel into the air, it is finished. These words from the Bible created our four seasons: thunder, lightning, and extreme weather. So, using our words today to balance the extreme weather will outsmart the evil corruption against others. As scientists discover more about what is out in outer space, there are more planets on which we can live. But wouldn't it be logical to rotate the weather around them first? Before they become livable. If it is possible to stretch out between each plant, the energy force that keeps the seawater down on one planet, if our words are around each planet in clouds rings. And tension forces. We rotated and used our wordsto take tension away, and the sea level spilt around other planets, as in past times. The possibility is that we have lived on these planets before. This is based on theory -insight and vision of past and present words. that create the clouds. If the earth had been flat before we broke the universe apart and had begun spinning around an axle, we could have become flat again. These wise words in the bible verse could save us from the worst happening flat; this solves many unanswered questions about rising water and when and if the planet stops spinning on its axle. What then. Do we give up? No, we found a way to create a balance towards our close planets. Mars, Mercury, Venus. Earth. by unravelling our past and present words and understanding their wisdom after we correct all past mistakes against others. We can balance the weather (called mother nature). Once we leave the tension words in cloud forms, we stay on a level with all cultures. Family tree first. Then, up one step time by lifting our word around these other planets and letting the seas flow around them, like at the beginning time, by lifting the tension words and images further out, we will be able to let the rising seas on Earth flow through, around create growing plant trees, hoping the weather above will make the rain, wind, sun, as on earth. Every idea, theory, insight, wonder, and vision will play a part in unravelling the real reason for our being. Our words and action deeds all form images in the clouds. Of the spinning planets.my foresight of how it might be. planets on which we can live. But wouldn't it be logical to rotate the weather around the planets first Before they become livable? If it is possible to stretch out between each plant, the energy force that keeps the seawater down on one planet, if our words are around each planet in clouds rings. And tension forces. We rotated and used our wordsto take tension away, and the sea level spilt all around other planets, as it had been in past times. The possibility we lived on these planets before is based on theory -insight and vision of past and present words- that create the clouds. If the earth is flat, this solves many unanswered questions about rising water and when and if the planet stops spinning on its axle. What then. Do we give up? No, we found a way to create a balance towards our close planets. Mars, Mercury, Venus. Earth. by unravelling our past and present words and understanding their wisdom after we correct all past mistakes against others. We can balance the weather (called mother nature). Once we leave the tension, words will be in cloud forms, and we will stay on a level with all cultures. Family tree first. Then, Around them, like at the beginning time, by lifting the tension words and images further out, we will be able to let the rising seas on Earth flow through, create growing plant trees, hoping the weather above will make the rain, wind, sun, as on earth. Every idea, theory, insight, wonder, and vision will play a part in unravelling the real reason for our being. Our words and action deeds all form images in the clouds.

CHAPTER 14

AFAR SIGHTED THEORIES, IN MY MIND EYES

WHAT IS THE PURPOSE OF using the bible verses to unravel the universe and everyday education? We all get parts of the bible coming in on us nationally. And these verses, our minds understand . can cause tension, accidents, and retaliation against others—the bible verses. Make understanding easier. Then, understanding different people's sfar-sighted theories. All theories will be helpful in time. Could it be, or have we had it all before? We made a mistake and broke the planets apart thus far. My theory of wonder and thoughts vision is that once we collect all our words, past and present, around all cultures, for levels of understanding wise wisdom words, check out the verse in the bible by using the alphabet and numerology. Break each down number, the Bible check verse that suits one family tree or wise word understanding levels. If we understand our words and have corrected our mistakes against others, we don't. Know their wisdom on all cultural levels, but only in our family tree. We could open passages around our nearest planets. Mar Venus mercury up to Jupiter. Has earth accumulated all the water as the tension of our words, anger, stress wars? All are in cloud form, some in images. Can the seas flow again around these planets if we use all these words, past and present? But would we reach a standstill in the universe, so would our movement have to be on levels to maintain a balance? If we stop spinning around the universe.

WORDS=W-23=O15=R18 D-4=S-19=79-16-7(79) PSALMS: Your land has been conquered. In this heathen nation, our lives have been defiled and are in heaps of ruin. This is why we need to act now and act fast on the levels of our word and wisdom. If the planet stops rotating around the axle, we will not have a balance. Repeating the same mistakes against others,=. As my wonder grew, I realised there had to be more to our lives than what we were experiencing. I heard a lot. We grew up hearing these wise old sayings. What was the reasoning behind them? That being said, there just had to be a purpose. With a lot of trial and error over the years and putting it all together, I found the way. The repetition of words in life will happen over the next generation. The same old things occur time after time and again in a generation if not corrected. The messages of wisdom and unspoken words came to me. I put the pieces together in my family tree and my mind. The more I mused, the more I thought there had to be a way. As I said before, I probably repeat sentences in chapters, but that's for each time I improve or add more to the sentence. The moral behind this book is to find an answer to life's most complex problems. A man has been looking for an answer for centuries, to no avail. But one way has been found: to the real reason for our being, how to proceed with time travel. We found a way to help us balance Mother Nature. The devil got outsmarted by god, and ideas to use our words to balance the weather cloud forms, all earth creation, forms images in the cloud images. Of words, actions, thoughts, and wonder. In-universe. Past words start to add up to the reality of our lives. And why God made us in his image to help with his plan; nursery rhymes, songs, and past sayings—vision, wonder—all play a part. When our forefather and mother started a sentence, it was interrupted, and then they would forget to rest and say what they wanted to say, so we only got part of the story they wanted to tell others. There is an adult meaning to all nursery rhymes. To keep children's minds down to levels they can understand, they lose concentration on their sentences once interrupted. And the thoughts about what would be said got lost in the interruption, and only half the sentence would get said. These half-sentences are called unspoken words, past and present. So, to unravel the whole picture and add the last sentence to the future, we return and start again. We were old when we were born. We grow, learn, make mistakes, and learn from them. Pick up pieces on the way, back up through our lives to age retirement. I may have repeated myself in this book, but this is normal, as some sentences could come back that need improving further. I have repeated some chapters from my first two books in this book. LIFE MOST COMPLEX PROBLEMS, AUTHER BETSYBOO, and LIFE BIGGEST JIGSAW PUZZLE. AUTHOR Elizabeth Cargill Campbell. as each one leads into the next for improvement. My starter books now have my full vision of theories: to help others improve further on the grounds of the safety of this planet and all artificial cultures. One thought when I was a youngster. Why were some children white, and others had dark cultures? As I went to school with different cultures at that age, I could understand this, so it got put back in my memory bank until I was ready to find an answer. Well, I was forty years old when God sent me an answer. The answer came thick and fast. Please accept my beliefs. I do know how we became different cultures.

Every chapter can tell a story about the representative verses in the Bible. Write down your thoughts and wonder about words that come back. Also, put the date when one first learned of theories and wonders. Vision words: as with all of us, only parts of the sentence come in. at any given time. Write down the time, date, and year of the acknowledgement. of any given situation idea. Words of forethought to further time. One person might know something early but can improve once it becomes known to another. Our words are the most essential part of this era of our lives. As we go, we balance out the weather. (Mother Nature) When Mother Nature helps balance our words, it guides us back on the right path if we stray too far in the wrong direction.

CHAPTER 15

How It Can Be Done: LISTEN WITHOUT INTERRUPTING

E ARE BORN OLD; WE grow in mind, body, and soul. Our mistakes as we grow up are the key to understanding our words and their uses to help balance out clouds. Our forefathers and mothers, subconscious minds, know the truth, book, got it wrong. How many times have we listened to others talk? How many times do we get interrupted before they say the whole sentence? This is how the catchphrase, or old-time saying, originated, as one starts saying something and the rest disappears from them. Where does the rest go? I wonder as we all do as youngsters. We wonder. We think. Often, our vision will see things the mind doesn't understand. So, it gets put aside, and one thinks about returning to those thoughts another day. But as it happens, the day never comes. When we retire, we forget what we put out there that we wanted to know more about our wonders, thoughts, and dreams. In our retirement, the best time to ponder past events is to correct all mistakes made by others. We don't need to look for our mistakes; they are all around our bodies. Just wait, and they will surface; we might see someone do something similar. Or rethink our lives. If we correct past mistakes as they come in using later years, we understand the wisdom of these mistakes in words. The old saying; Wherever one may be, let your wisdom go free says that wisdom is easy to carry but difficult to gather. How true this word past is, and how enlightening it is. Forfutures to improve further. The saying of words is easy to carry. To tell the story, they cling to the waist like a belt. May I ask what the word =still means in any language or Bible verse? 60-6- (6) The lamb broke the scroll's seal. I began to unroll the scroll. The first was a white horse of daylight colour. The second was a red horse. The third was a black horse. 4th was a pale horse, which means death. 5th horse hell. Who was given control? One-quarter of the earth. In the sixth seal of vast earthquakes, the sun became dark, and the moon blood-red in the eclipse. We have just experienced it on earth in 2024, meaning these six horses all represent colour. And different cultures are alive today. With respect, our words are so exciting and enlightening.

GIVE WITHOUT SPARING AND WANTING

The white horse is truth and honesty within ourselves. Every child born can follow one parent's wise words, or the other, in their side family wisdom of words action, of the family tree. Some could handle both sides, father and mother's wisdom from their family tree. Grow and learn further. I also put these phrases in this book, as it shows how the values we taught in the home all match up to verses in the bible. One way later to find and understand what each word means in God's true path to follow—our words your way and empty seven. If these plagues surface in the public arena again, we may have stopped learning and improving on the next level of our words to understand wisdom. After all, once mistakes are corrected, individuals are against others; this is the key to future levels: correcting past and present mistakes. We open up further energy forces and spread wise wisdom around all cultures to improve further or to hold (patience)

Don't stand behind me; I may not lead. Please don't stand before me; I may not follow, but stand beside me and be my friend.

Water sign = cancer, Scorpio Pisces RAIN TEAR DROPS,

Earth sign, = Taurus, Virgo, Capricorn KEEP OUR FEET FIRMLY on the ground

Air sign = Libra Aquarius, Gemini. LET THE AIR THROUGH

Fire sign= Leo. Aries, Sagittarius BRINGS THE WARMTH

I repeat these phrases in all my books. They are essential for our growth, in mind, body, and soul, and they are all so necessary for growth into our future to understand and stay on our rightful path, survival of human life, and this planet using our words. We all have negative and positiveness. Once we know the negative part of our lives, we can open the doors to our wise words. To explain these actions in our time,

THE FIRST STEP: We are born old and Learn to talk verbally

SECOND STEP: To live by the values taught in the home

THIRD STEP: Wait to touch what's yours before your time.

FORTH STEP: Correct all past mistakes against others.

FIFTH STEP: Understand the wisdom of those mistakes.

SIXTH Stay on level understanding first.

SEVEN STEPS, What for god to send next level wise words.

EIGHT-STEP God number.

NINE STEPS Grow in body, mind and soul,

TEN STEPS: Balance out mother nature with our words. Use our words to balance the weather. There are two meanings to all passages in time. A verbal one and a spiritual word. ones called (unspoken words)

Our words have no wings but fly for miles—by phone call or what we carry around our waist like a belt. Life is most tragic. We get old too soon and wise too late to see how our words match Bible verses. This is just one way to unravel the real reason for our being. Another person might find another way—that long-standing and long-lasting reason for our survival on Earth in human form. I repeat myself from my other two books in this part, as improvement comes with every sentence. As people need to understand, the stepping stones of my books can only be improved further as I became aware of the shortcut in science that involves the real reason for our being. As Winston Churchill said, "A nation who forgets its past has no future."

CHAPTER 16

FORGIVE WITHOUT COMPLAINING

AS THE HOROSCOPE SAYS, THE good and evil in us all are the negative and positive of one nature.

NEGATIVE AND POSITIVENESS.

To bring the horoscope into this learning process, We all have a negative side and a positive side; once we understand the positive side and separate the negative from our thoughts and actions, we all have a great personality created to handle every situation. This personality and the opposite of our signs create a whole nature. We should wait for the right eternal partner because of past mistakes. They are kept on a partner until corrected, and their wisdom is understood. One's past can't get through to upset our feelings thus far. In part of the horoscope, I have laid out signs opposite to what I've learned myself thus far. Each one has a role in our lives and the energy force of the sky. Fire signs bring heat and sun warmth, and Water signs bring rainstorm drizzle. Earth sign keep the evenness of our feet firm on the ground. Air signs bring in fresh, cold air around people. We would only move when it was our level, too. With our opposites above us so as not to go to extremes in any one direction, Relationships are built on trust. There is a big difference between just having a conversation with the opposite sex or giving another the come-on.THE 7 STARS, 7 GODS, 7 ANGELS, 7 PLANET 4 -7,s =28-10-1(10) Now we saw another angel coming down from heaven, surrounded by a cloud. With a rainbow over his head, his face shone like the sun, and his feet flashed with fire. He held in his hand a small scroll; he set his right foot on the sea, his left foot on earth. And seven thundercrashed their reply. Was this small scroll carrying the message to use our words to balance out the weather for generations to come? (Mother Nature) 3 seven 21 The key is the door to understanding the real reason for our being. Theories: We lived before, made mistakes, and broke the universe apart four times, each time having to start from scratch again. Hopefully, people will notice how meaningful our lives are this time and help balance out weather patterns, Mother Nature, and the level of one family tree first. Then, our words will be used to advance weather patterns. God has seven stars in the sky, but one has never been seen from Earth; I believe it is the star standing beside god to spread his words of wisdom ahead of us. Being sent down to the pits of hell to help others so they too can follow the right way to level others can understand, once all past mistakes are corrected back to the person done to other families. Lots of us sacrificed material possessions to understand the real reason for our being. Man has been looking for a way to stay alive in human form for centuries but to no avail. Only to have what we have done thus far wasn't good enough to keep human life alive; my theory on this is, and what I have seen by the vision, in past times, we made a mistake, and the elements of the universe suffocated earth, we had to start from scratch four times again in all. (About) This is one way it can be done in this day and age. Seem to forget that the almighty god made us in his image to help him with his plan to conquer the whole universe (REMEMBER), as we have had it all before. Once we make a mistake and break the energy force between each planet apart. And we had to start from scratch again. Christianity is based on the values we were taught as children; we had to live by them; the gods say to return to what one first knew and start again. Hemeans birth; as individuals, we are old, come back up through our life span and correct all mistakes against others back to the person they did against. Some of us sacrifice material possessions to understand the real reason for our being. Man has been looking for a way to stay alive in human form for centuries but to no avail. Only to have what we have done thus far wasn't good enough to keep human life alive; my theory on this is, and what I've seen by the vision, in past times, we made a mistake, and the elements universe suffocated Earth; we had to start from scratch again, four times in all. (About)This is one way it can be done, in fairness to all cultures of humanity. Each time, we had to start from scratch. Christianity is based on the values we were taught as children; we had to live by them. God says to go back to what one first knew and start again. Hemeans birth, as individuals, we are old, come back up through our life span and correct all mistakes against others back to the person it has been done to, or against

CHAPTER 17

THE WALKING STITCH 17: THE MATERIAL EARTH ACTION OF HUMANITY

AST, PRESENT AND FUTURE: A time to laugh, a time to cry, a time to work, a time to play, a time to rest, and sleep; this part is based on the theory of our actions all in one place when we have actions like exciting football games that extract all the public attention. Worldwide, everyone's attention is focused on one area. Could the axle get off balance and create an earthquake in the universe? I believe our energy force and word, action, spins us around on an axel past and present, and when all human life is focused in one era of the world, it very well could offset the axel; they say the axel spinning faster, not than before, but it could it be because we are getting the pathway right. This is based on theory, Science, research, time saying, song nursery rhymes, and vision of what is ahead. If the axle should come to a standstill, stop spinning like a golf ball stuck in a ruck of words. Are we prepared to handle the atmospheric conditions that will arrive on us? The energy forces the past and present that keep us spinning. So, by working out on level all cultures, the wisdom of our words that God intends us to use this time. God, with our help, will outsmart the devil. All his evil, harmful, or good words were found to be beneficial for them and will keep Earth safe from harm. But God needs all humanity to help. As the saying goes, God made us in his image to help him with his plans to conquer the whole universe. REMEMBER, this book explains one way it can be done in fairness to all humanity and cultures. On level in family trees first, then wait for messages from God to step in words, we can begin to understand the path god has in store for us. All cultures branch out like a tree on its own culture first,

OUR WORDS: What are the theories About the black hole forming in the sun? Vision and wonder about what will happen as it rolls on. Culture is going up through the sun and splitting the sun into four quarters, causing blackness in the universe. But was it a future vision of when the black hole appears from Earth? If a culture goes through the sun, each side human words life and actions. A black hole joins each culture in the middle. Face looking out into the family tree, would the one big sun fit into our miniature suns to warm part of the earth down the cultures? The Big Apple was said to be split into four quarters. Is it a good action or a dangerous time for the Earth? Using our words to balance out the weather and energy force worldwide will save any destruction to our planet. I saw a vision years ago of 4 suns out there surrounding the universe. Is there any scientific evidence to prove this is accurate? Material possessions and spiritual words all connect and leave images in the atmosphere. In cloud form, in our words, four suns and four quarters in the universe of the sun. Here on earth, families have four sons—the unspoken words in the black hole of cultures. Back-to-packages of the eclipses in April 2024, did God's hand pop out of the sun; by vision, I have seen four suns out in different parts of the universe. If one has experienced anything along these lines by vision, write down as one day they are needed for one reason or another.

BRIDGE OVER TROUBLED WATER our words.

It takes one step toward improvement on the next level of all cultures—three steps to heaven gates. The door is always open to those who correct mistakes as they go. Walk on the wild side. It depends on one's walking stick, not on others. The white horse is the truth, the light, and the way; first, be honest with yourself. Live by values taught in the home. Use patience in everything one does. Stop, look and listen to the same step we take to cross the road. Material possession, and other, is the spiritual way, unspoken words. If they say they found another planet like Earth out of the universe, there would be no life or trees. This should be true. How did it get there, and why are questions to be answered here? With theories, research, and parent-wise words, we can discover the reason for our being but stay on the level until it is time to take a step ahead. Is it just an image or reflection of the earth before humans got their hands on it? Or does it exist? I might add a couple of theories: I lived on this planet before, and by going too far too soon in the wrong direction, we took the light out of the day, and the planet got stuck into the black, dark universe. As will happen in today's life, we go too far off our levels and go in the wrong direction. Now, the other reason could be a planet in waiting as under the word, FLAGS (MEANS ALLCULTURES) (46=10=1) verse God is our help in times of trouble, use our word to help balance the weather patterns on a level all cultures as we need not fear if the world blows up. Could scientists see this planet as a stepping stone over when humans destroy each other? This word matches the bible verses and what they could mean for our future travels in the time of learning. We can combine the pieces by forming theories about what our vision sees or wonders about. There always has to be work to be done. NEVER TAKE life for granted, as we don't know what is around the corner, but by keeping an open mind on the ways of the world, we can be prepared for the unexpected; as we are taught as children go has a big book in the sky universe, of all we think say do, We can only see as far as the mistakes we laid on other, or the ones clinging to our waist like a belt. Understanding the way to use our words, good or bad, comes to us from our father and mother's parents, parents, brothers, and sisters. When we break words down, check the Bible verse to give one a clear vision of what words mean. When given an updated English bible, I first stumbled upon this idea to get a better meaning to reason for our lives. Another bible might have different verses with the same meaning but just worded differently from the older times. It's exciting and enlightening and is the reason for our being. So, let's try these words: PARROTS. Researched, and some live till 80- odds years old. Having a parrot in the home repeats itself with our words, and humans repeat the same mistakes they made before without

PARROTS= (17) Other fools are ill because of sinful ways. Their appetite was gone, and death was near. They cried lord in their troubled time, helped them and delivered them. he spoke, they healed, snatched from the door of death; my theory is that this verse explains what happens if we stop learning about the real reason for our being when we retire age, we give up as word keep changing, and elderly one in retirement forget the wise words they sowed. Ill health sets in, and the mind gives up. So, by writing down whatever comes to a head, good or bad words, breaking down into numerology and checking whatever verse in any chapter or verse in the bible, the bible does have all our answers. If one takes the word=psalms, LIE=26-8), Dismiss all charges against me, for I have tried to keep your laws and trusted you without waving.(8) I wash my hands to prove my innocence. As I see it, when we try to understand the real reason for our being, we try to understand everything at once, but it is best to consider what one needs to know at any given time. God won't send one answer to one wonder or thought from birth. He waits until the time comes when one's mind is ready to understand and improve.

CHAPTER 18

HOW TO READ THE UNIVERSE.

B Y KEEPING AN OPEN MIND to other theories, words, reasons, and ideas, as in time, all fit somewhere to help balance the weather patterns. Once a person has corrected all past mistakes from birth to adulthood against another, the path is straightforward to follow and understand the world's way. And there is a reason for our being. QUOTE: the bible states this: God said to go back to what one first knew them to come back to birth; we are born old, we stand up and fall, life like that. We think ahead, but our conscious minds don't understand or remember what the subconscious mind knows from past lives we lived: the wisdom of words until we at the level come in at any given time by applying the values and living by them, as taught by parents and guardians. We are kept safe, and messages from gods come quickly. In our lives, we go to Jesus first once we correct all past mistakes and go up to the gods for the rest of the time when travelling. Once all past mistakes are corrected, the pathway is clear to see by a vision of the future; one can't fix another past mistake. Otherwise, their mistake will be blocked one way. If, in the spiritual world, one stays on the same pattern path as they do every day on earth, then we have good coverage of how to keep the balance of the weather ahead. Hope is being able to see the light in all this darkness. It might have been the rolling clouds that gathered no moss; only the cloud would gather images and words to create the weather. Cling together around our waist and in cloud form in the sky. So, what does the phrase MIRE mean? Is the devil hole of wrongdoings?

MIRE -M-13-==I=9 ==R -18 ===E=5==MIRE 45= 9 PSALMS My heart is overflowing with beautiful thoughts. I will write a lovely poem. (9) revelations=And the fifth angel blew her trumpet and saw one who had fallen to earth from heaven and was given the key to the bottomless pit ==M I R E men inner. Ruler eternities. ==move inner ruler east.. == men inner raise eternal. One can see how one word can be used to explain how to bring in the lord's ways of letters. The word in eternal space, and how they can show the four seasons.

CHAPTER 19

The World Eclipses: KNUCKLES

I'LL FULLY EXPLAIN WHAT I write in this book. To the best of my ability, word =knuckles 94-13-4 plasms (94) lord god, to whom your vengeance belongs, Let Your glory shine out. Arise, judge, the earth. (13) he helps us by pushing us to make us follow his path(4) hear their insolence and see their arrogance. How they boast, elevations - (13) and now in my vision I saw a strange creature, rising from the sea with seven heads- ten horns ten crowns upon his horns,(4) there as I looked I saw a door standing open in heavens same voice sounded like a trumpet blast and said. COME:? I will show you what must happen in the future. We must use our words to Balance the weather patterns. God outsmarted the devil this time. All our words, good and bad, can be used on a level in all cultures regarding the ground survival of human life and this planet. Our words are the next step to understanding the wise wisdom of every Word. They match the verses in the Bible and tell precisely what one is looking for in an Answer..

THE BLACK HOLE IN UNIVERSE

It could be the unspoken word of a culture that can be pronounced in words of conversation. Based on theory, As God's word says in a verse in the bible. Return to what one first knew and then return to him (god). What he means by this is to go back to the beginning of BIRTH and come back up one's own life one lived from birth to adulthood. Correct all mistakes, big or small, when they return at one, back to the person who said against it or done it, too. Then, understand the wisdom of that mistake. One word in the verse of the bible can be explained a lot by using the alphabet and numerology. As shown in this book, then checking the verse in the bible, it Is unbelievable how our words match up with verses that explain in detail a meaning that improves on what we knew thus far; it's a short, quick way to unravel the reason for our being . as always been state. God made us in his image to help him with his plan. And each time we got this far, he would have ready the next step to improve further by balancing the weather on a level all cultures. We help save this planet and humans. Each culture on the proper levels can keep the weather at bay, but we also must have a man-woman level one up and one down without touching another level before one time and staying in one own family tree before going to the next level. Give our next generations something to live for, not die for. Let them feel they are part of improving our futures when we clean our bodies, not our minds. We leave our mistakes to others. But any individual can open the brown door into heaven until all errors are corrected against others. This book Outlines how it can be done on a level for all cultures. As we go, we balance out the weather, FIRE. SUN, WIND, RAIN, SNOW. THE FOUR SEASONS. The four angels at the corner of the earth help keep the balance F, S, W, R,=Future sun wind ruler or words ruler-future sun. Take note of nursery rhymes, old-time sayings, songs, thoughts, and wonders. Vision and have one own theories on them as these theories could help improve further ahead in time.

CHAPTER 20

THE WATERWAYS

Listen without interrupting, WORD DROPS

*O*F THE SPINNING PLANETS.MY FORESIGHT of how it might be. Always keep an open mind to their theories or ideas, as there could be something in their sentence one may be able to improve further,

THE SONG= RAINDROPS KEEP FALLING ON MY HEAD or could it be word drops, one drops a word to another person for further improvements, Raindrops and teardrops The most common thing that happens in our daily lives today, But in the reality of words is the raindrop; raindrops could have in the past meant WORDDROPS ON MY HEAD and how we send our wise words over to the next level improved on and back again

WORDDROPS.=132=16=7=(132-)lord, remember when my heart was in turmoil; I couldn't rest, sleep, or think there had to be more to our lives. And a way to use our words, good or bad, to keep the balance of weather patterns. In the past, the ark was Built to protect it from extreme weather. Nowadays, another way is to use our words, good or bad, to balance the weather on a level appropriate for all cultures worldwide. Will it be possible to spread the rising seas on Earth around another planet as we use up the words wisdom tension between us and the other planets? Mars, Venus, Mercury, and Earth only have life on them. If the seas were around these planets before we broke the universe apart, seawater would have blocked the flow of seawater only around Earth. .tis is a theory worth looking at or thinking about as if my theory is correct and the ring around other planets is whether partly done, the by unravelling our wise words we could put weather back around other planets as was.= revelations(16) and then heard a mighty voice. Shouting from the temple to seven angels. Now go your ways and empty seven flasks of the wrath of god upon the earth. The first angel poured her flask over the planet, and sores broke out.2nd angel poured out over the ocean. It became like water, and all in the sea died. 3rd angel upon rivers springs you are just in sending this, judgement, the fourth angel upon the sun causing it to scorch all with its fire. The fifth angel kingdom was plagued with darkness, and the sixth angel great river dried up. The seventh angel blew smoke into the air, and it was finished. These words from the bible created our four seasons: thunder, lightning, and extreme weather. So, using our words today to balance the extreme weather will outsmart the devil's corruption against others. As scientists discover more about what is out in outer space, we could live on more planets. But would it be logical to rotate the weather around them first? Before they become livable. If it is possible to stretch out between each plant, the energy force that keeps the seawater down on one planet, if our words are around each planet in clouds rings. And tension forces, and we rotated, by using our words to take tension away and sea level split all around other planets as in past times. The possibility we lived on these planets before is based on theory -insight and vision of past and present words. that create the clouds. If the earth is flat, this will solve many unanswered questions about rising water when and if the planet stops spinning on its axle. What then. Do we give up? No, we found a way to create a balance towards our close planets. Mars, Mercury, Venus. Earth. by unravelling our past and present words and understanding their wisdom after we correct all past mistakes against others. We can balance weather (mother nature) once we are left with the tension of words in cloud forms that stay on a level in all cultures. Family tree first. Then, up one step time by lifting our words around these other planets and letting the seas flow around them, like at the beginning time, by lifting the tension words and images further out, we will be able to let the rising seas on Earth flow through, around create growing plant trees, hoping the weather above will make the rain, wind, sun, as on earth. Every idea, theory, insight, wonder, and vision will play a part in unravelling the real reason for our being. Our words and action deeds all form images in the clouds. One must understand that theories, ideas, visions, and insights are my own based on old times saying nursery rhymes, songs, and parent-wise word wisdom. It was found IN The middle of the moon. WATER. Is it salt or fresh remains to be seen? This is worth more theories. The moon follows as the saying goes, There and back, to see how far our words go. Moonlight, starlight, how I wish on falling star tonight. FALLING STARS TONIGHT,=m-13-015- 015_ 14N- MOON--Then angel, standing portraying things, comes with the sun and moon beneath her feet

MOON = WEATHER 13 =15= 15= 14= move over on north. or men on over next. This is sent to you from the one with sevenfold spirit. Gods and seven angels. There is a part in the bible where Jesus raises his hand with the power of thought of words. He raised the sea. Could our world increase soon from around the universe, opening up the seawater from the earth around this other planet? How could it be before we made a mistake and broke the universe apart? After inside Mars, Jesus raised water, and the ice is melting on Earth, which will cause the sea level to rise; this excess seawater could sprinkle around these other planets once our words, wars, and tension being lifted may soak up the cities that got buried before in time. I wonder.

CHAPTER 21

AFAR SIGHTED THEORIES IN MY MIND EYES THE WAY OF OUR WISE WORD. FOR FUTURE PROGRESS.

*F*AR-SIGHTED THEORIES. COULD IT BE, or have we had it all before? And made a mistake and broke the planets apart thus far. My theory of wonder and thoughts vision is that once we collect all our words, past and present, around all cultures, For levels of understanding wise wisdom words, check out the verse in the bible by using the alphabet and numerology. Break each number down; the Bible check verse that suits one's family tree or wise word understanding levels. If we understand our words and have corrected our mistakes against others, then understand their wisdom on all cultural levels, we could open up passage around our nearest planets. Mar, Venus, Mercury, and Jupiter have all the water accumulated on earth as the tension of our words, anger, and stress wars. All in cloud form, some images. Can the seas flow again around these planets if we share all these words, past and present? But would we reach a standstill in the universe, so our movement would have to be on levels to keep the balance?

Of the spinning planets. My foresight of how it might be.

THE SONG= RAINDROPS KEEP FALLING ON MY HEAD, but were they originally word drops that fell on our heads to remind us how to use our words in the future, to balance out Mother Nature (weather patterns)? Raindrops and teardrops are the most common things in our daily lives today. Still, if words were raindrops, they could have meaning.

WORD DROPS ON MY HEAD: How we sent our wise words over to the next level gets improved and back again. Words on ruler distance and Earth only have life on them. If the seas were around these planets before we broke the universe apart, and all the seawater surrounded Earth, This is a theory worth considering if my theory is correct. The ring around other planets is that the weather is already partly done from the past. Then, by unravelling our wise words, we could put weather back around other planets as was. as

REVELATION (16). Then, we heard a mighty voice. That was were shouting from the temple to seven angels. Now go your ways and empty seven flasks of the wrath of god upon the earth. The first angel poured her flask over the planet, and sores broke out.2nd angel poured over the ocean; it became like water blood, and all in the sea died. 3rd angel upon rivers springs you are just in sending this, judgment, the fourth angel upon the sun causing it to scorch all with its fire. The fifth angel kingdom was plagued with darkness, and the sixth angel great river dried up. Seventh angel into the air, it is finished. These words from the bible created our four seasons: thunder, lightning, and extreme weather. So, using our words today to balance the extreme weather will outsmart the devil's corruption against others. Word drops (wisdom on ruler distance. THE BLACK HOLE. I wonder. As scientists discover more about what is out in outer space, we could live on more planets. But would it be logical to rotate the weather around them first? Before they become livable. If it is possible to stretch out between each plant, the energy force that keeps the seawater down on one planet, if our words are around each planet in clouds rings. And tension forces. We rotated and used our words to take tension away, and the sea level split around other planets, as has been the case in past times. The possibility we lived on these planets before is based on theory -insight and vision of past and present words. that create the clouds. If the earth is flat, this will solve many unanswered questions about rising water when and if the planet stops spinning on its axis. What then. Do we give up? No, we found a way to balance towards our close planets. Mars, Mercury, Venus. Earth. By unravelling our past and present and understanding their wisdom after we correct all past mistakes against others, We can balance the weather (called mother nature). Once we leave the tension words in cloud forms, we stay on a level with all cultures. Family tree first. Then, up one step time by lifting our word around these other planets and letting the seas flow around them, like at the beginning time, by lifting the tension words and images further out, we will be able to let the rising seas on Earth flow through, around create growing plant trees, hoping the weather above will make the rain, wind, sun, as on earth. Every idea, theory, insight, wonder, and vision will play a part in unravelling the real reason for our being. Our words and action deeds all form images in the clouds.

CHAPTER 22

A FAR SIGHTED THEORIES IN MY MIND EYES: It is all in the eye of the beholder.

THE EYES HAVE IT: VISION, wonder, thoughts, and insight. Since we were children, we have all wondered what is out there. But never follow through as our education school work takes over. but when we retire, this time to reflect on what we wonder about in the sky universe. Far-sighted theories. Could it be, or have we had it all before? Past humans made mistakes. And we have broken the planets apart thus far. My theory of wonder and thoughts vision is once we collect all our words, past and present, around all cultures. Of levels of understanding in wise, wise words. Check out the verse in the bible by using the alphabet and numerology. Break each word down, number, and then check the verse in the bible that suits one own family tree or wise word understanding levels. If we understand our words and have corrected our mistakes against others, then understand their wisdom on all cultural levels, we could open up passages around our nearest planets. Mar, Venus, mercury. Up to Jupiter. Has earth accumulated all the water as the tension of our words, anger, stress wars? All in cloud form, some images. Can the seas flow again around these planets if we part all these words, past and present? Bring down and understand the level of wisdom. But would we reach a standstill in the universe, so our movement would have to be on levels to keep the balance?

ONLY WISE WORDS, THE BLACK HOLES

WORDS=W-23=O15=R18 D-4=S-19=79-16-7(79) PSALMS: Your land has been conquered. In this heathen nation, our lives have been defiled and are in heaps of ruin.

=W, ISDOM O, N R, ULERS D, ISTANCE .S, UN. We want to show people how it can be done in all cultures. As we go, we keep a balance on weather patterns. If we had it all before, the answers are out there, waiting to be unravelled again. Do our words and reason cause life and death, and how can it be done? In fairness to all people and cultures. I always wondered if there was anything more to our lives then. What we believe is living. Little did I know that the Bible verses could be used to understand the real reason for our being. And the right passage will take time. Looking at the whole picture ahead and beyond, my path in time took a turn in a specific direction. I hadn't planned for my future. I wanted to save, travel, or maybe go into politics, as I grew up having political conversations in the house. But we had to find the answer to life's most complex problems somewhere along the line. But as a teenager, I changed the course of my plans immensely, especially for us, and I realised why God created us. I see friends and family all running around saying and doing the same things against each other, getting nowhere, and repeating. This generation just saying the word verbally without understanding the full potential. Of these uses., I realised there had to be more to our lives than what we were experiencing. I heard a lot. We grew up hearing these wise old sayings. What was the reasoning behind them? That being said, there just had to be a purpose. With a lot of trial and error over the years and putting it all together, I found the way. The repetition of words in life will happen over the next generation. The same old things occur time after time and again in a generation if not corrected. The messages of wisdom and unspoken words came to me. I put the pieces together in my family tree and my mind. The more I mused, the more I thought there had to be a way. As I said before, I probably repeat sentences in chapters each time I improve or add more to the sentence.

CHAPTER 23

PRAY WITHOUT CEASING, AND WITH MEANING

To LIVE BY THE VALUES taught in the home, too, comes from the verses in the bible. Put together in the black holes. When our words mount up and congregate together, humans can express them, so they gather together like solid rock. Come down underneath their words? that can't be spoken or said verbally? Is it possible that darkness congregates in a mass and binds together, creating a black hole? The different cultures around these words. This is a theory of my insight of merging starlight and darkness. Earth will be in darkness if we take the light out of the day. This is one way it can be done, in fairness to all humanity and cultures. If someone else finds another way, well and good, as man has been looking for an answer for centuries, to no avail, that carries through to everlasting life on this planet will be welcomed. All in all, we found a way to balance out the weather and prevent Noah's Ark days from happening again. In this book, I based my words on theories of my family's wisdom and phrases—wonders, thoughts, and ideas. As history says, man has been looking for a way to balance out the weather and proceed further in time if the axle should stop spinning. This is one way it can be done in fairness to all cultures of humanity—points of importance from the bible. God said to go out and multiply, which means have babies, as the more human life we have, the easier it would be to balance out our words and images in the cloud forms around Earth and the other planets; god created humans to help with his plan. Remember. Conquer the whole universe. Theories on what one might come to mind while doing everyday work, such as one day getting the answer of improvement, can also be taught in schools. Things come on children while in classrooms. Write down the words, good or bad, then check the verse in the bible and use the alphabet and numerology. The Bible states this. When we pass through life and correct our mistakes, they block our being and prevent us from moving further ahead in the wise words track of levels. As we go, we leave a level for the next generation to improve. Some people must act out the words that form in their person, while others understand what they are. Please write down the parts that interest you. One step at a time, on a level of all cultures, some of what I've experienced myself in life, others taken from verbal conversation. Unspoken words, theories, wonder, thoughts, vision- one mind's eye. As they say, it is all in the eyes of the beholder.

EYES = east yonder east south. or eternal years enter the sun. I have seen, by vision, four suns out in the universe. Was it past, or is it in the future,

EYES= 5-25-5 19-=5 4=(9) revelations nine= .(9) The fifth angel blew her trumpet, and I saw one who had fallen to the earth from heaven. To her, I was given the key to the bottomless pit of all past mistakes against each other. The sun and earth were darkened by smoke. With these words said, the universe is darkened by smoke, which may be the reason why smoking started. To live by the values taught in the home. Our past words and reasoning are out there, and I believe in cloud-form images. When reading this book, one must remember that based on my experience, wonder theories, vision, thoughts, and research, my wonder of what is out in the universe, I always knew there had to be more to our lives than the one we live today. Raindrops keep falling on my head, But that doesn't mean my eyes will soon turn red. Crying is not for me. Cause I'm never going to stop the rain by complaining Because m free Nothing worrying me be long till happiness steps up to greet me, song Raindrops keep falling on my head Crying not for me, because never going to stop the rain by complaining Because I'm free Nothing's worrying me, so true our word in song written by Bj, Thomas. Could this song come from pastwords? Word drops keep falling into my head; we will understand the following level over our heads to understand the real reason for our being. This theory is just a wonder. Could it be that way?

THUNDER= 20-8-21-14-4-5-18=(90=9) (Lord through all generations. You have been our home before the mountains were created. Before Earth formed, you were god without beginning or end. (9) revelations. The fifth angel blew her trumpet, who had fallen to earth from heaven. To her, she was given the key to the bottomless pit. THUNDER-=Time Heaven under the north distance enters rulers, or= Turn high under the next distance enters right. How can we balance out the weather with our words, good or bad,

THUNDER- t,h,u,n,e,r= t, time h, even, under n,orth eternity. Raise or T, ime h, even, u,per n, od d, own e, enter r, ruler.

WORD DROPS, same. Wisdom over raised distance, distant rulers over people sun.

RAINDROPS.114-15-6(114) From the land of foreign language tongue.Mind, body, heart, and soul. Grey sky days, grey skies not a breath, wind blowing tension hold clouds and word upright, they call elderly grey power. Those who learn from their mistakes then understand the wisdom of the world DROPS, and the same. Wisdom over raised distance, distant rulers over people sun.

CHAPTER 24

WORD DROPS, travellers' insight of unspoken words astral flying, vision all based on theories on how it might been,

TRAVELLER INSIGHT: WHEN ONE GOES astral flying, one should not make a sound unless, in other words, one looks, but one should not touch unless one reacts to help another whose life is in danger. One step time through the travels universe. If we had it all before, mistakes would have broken the universe apart thus far. Then, we have all the answers we are looking for and waiting to be unravelled. If we first stay in our family tree, stepping out on level one is best. Saying, wherever one may be, let your wise words go free. The reason is that this book enlightens us about what it is all about. This is one way that we can level out and balance the weather patterns; these kind words out there will be able to improve the down level of all cultures; if one should read my book on this, THOUGHT-SCAPE, On how we can balance the weather as we go, take from its parts one has wonder about themselves in past times of growing up. Man -Women one up, one down, then visa versa. Don't touch what's not yours before one time, in material possession and the unspoken words ahead. Yours before one time, the eagle eyes ahead in the spiritual world watching our every movement in time.

EAGLE: 30-3=(30)psalms)You restored my health. You brought me back from the brink of the grave—from death itself—and here I am, alive. (3) revelations. This message was sent to you from the one with the seven-fold spirit of god. And sevenstars. I have seen the eagle out there. I applied the actions of looking but not touching. So, in the future, in our words, that will carry weight for years to come, as I was ready for that part in time and have yet to understand it.

History talks about Odin losing one of his eyes in agreement to know the wisdom in life—past times—for a taste of water. This insight has brought a lot of meaning to words from all cultures, such as catchphrases, up and down all family trees, and even different cultures. All in the eye of the beholder. And The eyes have it in One's mind's eye. This known verse is in the bible written in proverbs, as it is accurate, this word originated from half sentences being said.

WORD EYE= 5-25-5. =35=8=8 =proverbs) Can you hear A voice standing at the heaven gates? And at every fork in the road. And at the door of every home. Listen to what is being said.

35-8 -8PSALMS. Let them be overtaken by sudden ruin, caught in their net, and destroyed. (8) Our majestic glory of your name fills the earth and overflows to the heavens. (8) Revelations, When the lamb had broken the seventh seal there was silence throughout all heaven I saw 7th angel stand before god and was given seven trumpets Over centuries down the family trees, this verse would form into time, saying IN THE EYE OF THE BEHOLDER, THE EYES HAVE IT. IN

ONE MIND, These old-time sayings originated from Odin's actions in the past, which gave his eye to understand more wisdom in time.

LISTEN WITHOUT INTERRUPTING

. THE LIVING BIBLE, any bible verses are good, as when reading them, enlighten one idea of the passage in time god intends us to take. People in glass houses shouldn't throw stones. The meaning of this catchphrase has changed over time as some people would accuse and blame others for things that were no business of theirs, and they get it all wrong, or they make the same mistake. These mistakes in accusing words get passed down through generations and come out in different forms around other people. That being said, our words and actions cling to our waist like a belt. The body heals once a mistake is made. And the wise wisdom is understood. These three main words can tell a lot of meanings in the bible,

GLASS =58-13-4=(13) And now, in my vision, I saw a strange creature rising out of the sea. It had seven heads, ten horns, and ten crowns upon its horns. (4) revelations. Then, as I looked, I saw the door standing open in heaven. And the same voice I heard before, that sounded like a mighty trumpet blast spoke to me: I'll show you what must happen in the future, the rainbow glowing. Arguing over foolish ideas, silly myths and legends; spend your time and energy keeping spiritually fit. Bodily exercise is all right. Spiritual exercise is much more critical. And a tonic for all you do. You know this not only in life but also in the next. Let them follow the way, with our words, to balance out weather patterns in cloud form and your clean thoughts.

STONES=92-11-2(11) psalms- You have made me strong as a bull in understanding our next move to balance out Mother Nature. Even in old age, they will produce fruit. Be vital and green.

(11reverlations,) Do not measure the outer courts. I was told that being turned over to the nations,

(2) this is writing to inform you of a message from him who walks among the people and holds their leaders in his right hand; this message is from him, who is the first and the last, who was dead and then came back to life. As one can see, these past verses explain how we arrived at most of our time by saying sentences that only got half told when trying to explain the meanings of verses in the excellent book. By checking out words that come in one us, good or bad, we can find the correct answer in the bible verse: stay in one family tree first. HOUSES=87-15-6=98=(15) revelations.= And then I saw another mighty pageant showing things to come. Seven angels were assigned to carry down to earth the seven plagues. The seventh poured his flask into the air, saying it was finished. Then thundercrashed and rolled, lightning flashed, and great earthquakes occurred.

(These twelve verses tell what putting pen to paper is. I watched as he broke the sixth seal. There were vast earthquakes, and the sun became dark like black cloth. The moon was blood red, and then the stars appeared to fall to earth like green fruit from a fig tree. Buffeted by mighty winds and starry heavens disappearing, this sixth verse in the bible could be the answer to how the heavens are out there at a distance, as well as the moon and the stars of different cultures. Words are the light between all this darkness,

The word glass, says it all.

GLASS -g, ods l, evels a, ll s, sunshine weather movement. GLASS-. Go left at Sunshine.

STONES. s, un t,ime o, ver n, od e, eternal s,hine.

Weather, movement, STONES, sun turn over north enter south.

HOUSES =high over upper sun eternal sun

weather movement, HOUSES. H,old o,n, u,pper S,outheast S,outh

G S H =7-19-8=34 7(34)psalm The earth belongs to god. Everything in the world is his. He pushed the oceans back to let dry land in (7) for the angels of the lord to guard and rescue all who reverence him.

GSH= God saves humans

GSH God, sun heaven.

GSH, go south heaven..

UP THE CREEK WITHOUT A PADDLE.

When the seventh angel blew the smoke into the air, was it partly polluted from p the space in heaven surrounding Earth? Could it be part of the pollution we are experiencing today on Earth? Was the air around the universe already polluted? Could that be what we call the ozone layer? There is much to understand and learn about the real reason for our being. Write down what one wonders about when young and what God answers. It will come, but when another story.

EAGLE EYE==45-9(45)PSALMS—I am as full of words as the speediest writer pouring out his story. Revelations(9) Then the fifth angel blew her trumpet, and I saw one who had fallen to earth from heaven, and to her was given the key to the bottomless pit. When he opened it, smoke poured out from some vast furnace. And the sun and air were darkened by the smoke. Could this smoke be part of the pollution surrounding the earth today? One of my wonders in time.

EAGLE-EYE= e,ternal. A, ll g, od l, levels e, enter e, eternal, y. ear. e, arly. Eagle. East at ground level, enter east years early, e, a, ll g, ods l, levels e,arly e,ast y,ear-e, nd. or something like that, depending on the level at the time, so one can see how important it is to live by the values taught at home. These words all came from Bible verses over time. The more I see the answers from the Bible verses, the more my hand becomes the speediest writer. It is so enlightening, engaging, and accessible.

CHAPTER 25

THE NURSERY RHYMES

ALL OLD-TIME NURSERY RHYMES WERE created to keep a child's mind down on a level they can understand, as there is a whole meaning in them all. Please put your meaning in the family tree first, as every meaning will have a place in time. What does the adult mean in this nursery rhyme song and old-time sayings? Nursery rhymes tell a story, too, one for children and one for adults, to improve further with theories.

(1) JACK AND JILL Went up the hill to fetch pail water. Adult means. Jack fell and broke his crown, and Jill came tumbling after him. This action in life by humans is every time our past comes back at us and is not corrected, and it causes us to unbalance and fall.

(2) LITTLE BOY BLUE. Blow your horn, sheep in the meadow, cows in corn, adult meaning. We are on different cultural levels—the wisdom of our words.

(3) Merry had a little lamb every that merry went the lamb sure to go. What is the adult version in our words and theories? This is the adult version. All cultures on levels of

understanding wisdom once all past mistakes are corrected; a little lamb could represent a child wondering what, how, and when.

(4) OLD MOTHER HUBBARD went to the cupboard to get the poor dog a bone. When she got there, the cupboard was bare, so the little doggy had none. Adult means when they rose to another level, the wisdom of our words wasn't done, so they lost their way,

(5) BA BA BLACK SHEEP, Have you any wool? Yes, sir, three bags full, one for the master, one for the dame, and one for the little boy who lives down the lane.= adults, meaning = man at the top, lady at the bottom, visa versa, total wise words, then down to little boy living down the road, our action in real life parts wisdom goes on in children. A child follows one family's wise words or understands both parent's and father's side of the family tree. Pastime nursery rhymes with adult meanings.on the way to go.

(6) HAY DIDDLE DIDDLE THE CAT FIDDLE, COW jumped over the moon. The little dog laughed to see a fun dish run away with the spoon. We would use the moon to send our wise words to the next level to understand and improve on, then back again. People used the moon on all planets to do this; the moon would light the way

(7) HICKORY DICKORY DOCK. The mouse ran up the clock, and the clock struck one. The mouse ran down. Time and clocks are critical to get the wise word rotating around all countries.

(8)   HUMPTY DUMPTY sat on a wall. Humpty Dumpty had a great fall, and all the king's horses and men couldn't put Humpty together again. Adult, meaning when we trip and fall, we break our bones. Our uncorrected mistakes are the same idea as when we lost our way in time and broke the wisdom of our words apart. And could find a way back until now, one way can be done. Apart.are causing the fall, blocking our way.

(9)   IT IS RAINING, POURING. THE OLD MAN IS SNORING: It's time for work, playtime, thinking, and rest. It is time to be patient.

(10) A Little Teapot Short and Stout. Tip me over and pour me out. Our mistakes cling to our waste as does our wisdom once understood, so tip me over, pour out the wise word for the next level

(11) WHEELS ON THE BUS GO ROUND AND ROUND; we go round and round the family tree. If our past mistakes are corrected, then we block ourselves from going further

(12) THE WISE OLD OWL SAT IN AN OAK. THE MORE HE MUSED, THE MORE HE HEARD THE values we taught at home. Look, but don't touch, ask, and one will receive,

(13) JACK AND THE BEAN STALK Jack climbed the bean stalk to be first ahead of others and he missed the levels in between so he fell to the ground and got hurt. So in life be patient on all levels of one family tree of wise words wisdom

OLD TIME SAYING, In the eye of the beholder. Listen Without Interrupting, Share Without Wanting and Listen Without Arguing— all these mistakes we make through life against another.

SONGS: Don't fence me in; teardrops keep falling on my head. Or were the word drops at the beginning? But in the future, will death, in our words, carry weight for years to come? This is true of life. These are just a few nursery rhymes with an adult meaning, like songs and catchphrases. Unspoken words need to be unravelled in adulthood on the level of fairness of all cultures. If the gods were humans like us and lived on earth in the past, their idea would be to create a plan for humans to stay alive in human form earth form until eternity. This is the same action that adults take to build their businesses. We all make plans of one sort or another. But the wise people's plan was for the survival of all human life to survive on Earth and any other planet.

CHAPTER 26

THE TONGUETHE BEGINNING OF TIME

Time to Nod under Eternity

I WILL WRITE COMPLETE VERSES FROM the English Bible that clearly describe the words used in everyday life and the values taught in the home. That was an old-time saying. If one pokes their tongue out, we are told, the wind changes, and one will stay like it. Another one was. one would ask, How old are you one would say. The answer would be As old as one TONGUE but slightly older than one tooth. This has been prevalent in past years: TONGUE could arise from the beginning when God's Time to nod under eternity. Little phrases were said to children at home by parents.t was a nice way to put things a child's mind understand

ALTHABET AND NUMEROLOGY, LETTER AND NUMBERS.

The word TONGUE 82-10-1(82) PSALMS: God stands up and opens the heavens courts; he pronounces judgment on the judges. They refuse to listen to evidence of how long they will shower special favours. On the wicked,(10) Lord, why are you standing aloft? And far away. Come and deal with all these proud and evil women and men who viciously provoke the truth. Revelations. (10) I saw an angel carrying down from heaven, surrounded by a cloud. With a rainbow over her head, her face shone like the sun; her feet flashed with fire, and she held open in his hand a small scroll. She set her left foot on the sea and her right foot on earth. Seven thunder crashes their reply. (1) This book unveils some of the future activities soon to occur; in the life of Jesus Christ permitted to reveal these visions, an angel was sent down from heaven to explain this wisdom meaning. Write down everything we heard say in the past; in the present, one wonders what the answer was kept secret. Could the words still need to be bespoken? This way, how can we use our words to balance out the weather? Are they the words that still need to be spoken from back then? That was kept silent? Are the words of our time, 20th century now, to be used? For us to understand and put it in place. WORDS that we spoke over the years, past and present. And words to come. God says the beginning, the end, the first, the last, the one who is, was and is coming again. I am the first, the last, the living one who died, who now lives again. And who has the keys? To hell and death. Be afraid. Write down what you have just seen and what will soon be shown. Seven stars, one saw in the right hand and seven golden candle sticks. These unspoken words are kept for this time in our lives to balance out the weather pattern as we go; our words form the clouds in all different cultures, which creates Mother Nature and the weather on all levels of cultures in fairness. The wise old owl sat in an oak. The less he spoke, the more he heard. How true in life. I might mention that I had an Asian friend say to me once that they were working on training at night and that it was OK during the day. I can see how that will benefit Humanity's everyday life.

HELL=8 -5 12-12=37 =37=10-1 Same as verse from tongue num 10

DEATH=4-5 -1 -20- 8=38=11 -2(11) Now I was given a measuring stitch and told to measure the temple of god but not measure the outer court as that being turned over to the nation, So the earth was turned over the country to help god with his plans, and that is to conquer the whole universe. Have we had life on these planets close to Jupiter before making a mistake and breaking the universe apart? I wonder. This book jumps around on different subjects: words and verses. Single letters. And insight into the meanings. Most common words are broken down in bible verses so that one can see the good in all our words. Help to improve further over time. The most common word used to match up the same verse is mostly as it opens the mind to the proper meanings. As we go, we lighten the load off our backs. For more of the next generation to improve further. It is also enjoyable.

HELL= heaven, eternal long life

DEATH=distance eternity all time heaven,

Weather. Hell. H, old e, ast l, ow. L e,ast a, ll t, urn h,igh.

This book is about how it can be done in fairness to all cultures and mankind. They say the answers are all in the Bible, but most people wonder where. This proves that our words will play a big part in balancing out the weather. As shown in my book, these verses. are In the Bible match words of everyday use. Our words can be used in two or three ways if they keep repeating repeatedly. We take the word, Sweetheart. These verses all come from my English living bible edition. All Bibles are good as they represent all cultures and family trees. Our values we taught at home come from the bible. word

SWEETHEART(124-25-7. (124) Psalms- If the lord had not been on our side, we would be swallowed up alive by our enemies, destroyed by their anger.

(25) Don't let my enemies succeed. (7)revelations) I saw four angels standing on the earth's four corners, holding back four winds, and the ocean became smooth as glass, as another angel coming from the east carrying the great seal of the living god Jesus.

WORD.SWEATHEART= S,UN W,IND E,TERMINAL A,LL T,IME - H, EAVEN ETERNAL ALL RISE TURN

SWEATHEART.= SOUTH WIND EAST ALL TURN HOLD EAST AT RIGHT TIME.

This could have been how it was done in past times, balancing the weather, but at this time, we have to use our words. If we made a mistake, we broke the universe apart. Our anger, phrases, and family tree words sent the seas down around Earth away from the other planets. Once we get our words back in place around all cultures on a level, the rising seas of ice melting on Earth will spread around MARS, Venus, and Mercury again. This paragraph is a repeat of another chapter, as more words are needed for lighting.

ONE MUST REMEMBER THIS BOOK IS BASED ON THEORIES OF LOGIC, VISION, and insight. Research and wonder about old-time nursery rhymes. And unspoken sentences The end of the bigging, the answers are all in the bible. Verses from every word of mouth of all cultures. This is one way it can be done in fairness to all mankind, by using words to understand the right path through time. As we go, we will balance the weather out. And prevent Noah's ark from happening again. Give others something to live for, not to die for. Learn something new every day. Grow in mind, body and soul. The end of the beginning. The clock face, The windmills, all action past times that showed the way of action and movement of our words.

MANY HANDS DO LIGHT WORK. IN THIS CASE, two humans on one level—a man and a woman—one up, one down, and vice-versa, to keep a balance on all levels with our words of wisdom. Some people can understand a word from a woman better than a man, or vice versa. I've put in the same chapter two or three times in this book. This is because I've added more wise words to it, which fall into the category. Once one reads it and believes they knew part of, then writes down what time, year, or month one understood the part, this puts on on levels with others who were aware than two, but the family tree may venture out in different direction through marriage. Or wise words from family.

BALANCE OF THE WEATHER

THE END OF BEGINNING

www.ingramcontent.com/pod-product-compliance
Lightning Source LLC
Chambersburg PA
CBHW020321130626
46549CB00003B/960